MIMESIS
INTERNATIONAL

PHILOSOPHY
n. 74

AF870877

Gregorio Tenti

THE PATHOS OF THE EARTH

A Critique of the Planetary Syntax

MIMESIS
INTERNATIONAL

© 2026—Mimesis International
www.mimesisinternational.com
e-mail: info@mimesisinternational.com

Isbn: 9788869775529
Book series: *Philosophy*, n. 74
Series Issn: 3103-3865

© MIM Edizioni Srl
P.I. C.F. 02419370305

Cover image: Paolo Soleri, *Arcology, City in the Image of Man 8*, Arcosanti, Arizona, USA © Cory Doctorow / Flickr – CC BY-SA 2.0

CONTENTS

ACKNOWLEDGEMENTS

With thanks to Andrea Bardin for the invaluable advice on chapter 1; to Francesco Sticchi for the enlightening Deleuzian exchanges; to Ernesto Giannoccaro for the countless Paretian coffees. And to Alice, earth-bound star-holder.

This book is a result of my Marie Skłodowska-Curie project at the Universitat Pompeu Fabra of Barcelona, titled *Understanding the Sixth Extinction as a Hermeneutical Challenge* and funded by the European Union's Horizon Europe research and innovation programme (Grant Agreement No. 101105444).

INTRODUCTION
CRITIQUE OF COSMIC REASON

> Ask yourself whether your question is not posed from a standpoint to which I cannot reply, because it is a perverse one.
>
> Karl Marx, *Economic and Philosophic Manuscripts of 1844*

> Who is able to put himself in this kind of 'standpoint of the cosmos in itself' and what could such a standpoint mean?
>
> Antonio Gramsci, *Prison Notebooks*

There exist many cosmic reasons, and many more have existed in the past. The question that animates this book is: what is *perverse*—in the sense mentioned by Marx—about the cosmic reason of Western modernity, and who takes on themselves the standpoint of the cosmos today?

The cosmic reason of Western modernity is a succession of totalities, or operations of totalization, aggregating different practices and domains of knowledge, according to different social-historical configurations. As complex aggregates of practices and data, cosmologies refer to reality and construct it in equal measure. They are in all respects *social axiontologies*, to use a Simondonian term: operations that define reality insofar as it exists both in itself *and* as a social fact, as the common horizon of different human standpoints. This axiontological element came to the fore when cosmology lost its scientific character during the 18th century. Since then, as clarified by Kant and later summarized by Wittgenstein, the world has become the "totality of facts" rather than the totality of things,[1] where "facts" are understood as the results of human doing (*verum esse ipsum factum*, said Vico: reality equals history). The world is what we make of it, and only in this sense can we talk of it in general. This fundamental shift corresponded to the emerging of the public sphere in Western societies and to the historical affirmation of European bourgeoisie.[2]

Cosmological totalizations are not pacific processes. Being social in nature, they unfold through contrast, conflict, asymmetric distributions and strategic relations; they may end in an act of pacification built on an apparently universal consensus only after they have established a historical hegemony. From a critical standpoint, it is therefore necessary to ask: who takes charge of constructing the absolute in which we live? This question is rendered necessary not by the cosmological fact itself, but by the naturalization of its operative and practical nature. By separating itself from the practices and subjects that enacted it and by hiding its material roots into a 'pure' discourse, totalization becomes sanctification. It is only at this stage that *das Ganze* turns into *das Wahre*, the whole becomes the truth, as Hegel had it, and that someone can object that, on the contrary, *das Ganze ist das Unwahre*, as Adorno did.

Thus, cosmology as I intend it includes the scientific discourse on the universe, but does not coincide with it. Science does provide a trustworthy access to reality, but 1) it does not exist in a vacuum and is always interrelated with praxis and culture, 2) it is a human (i.e., historical) mediation that constructs the way in which reality can and cannot be known and 3) it is not the only way to effectively relate to reality. The history of modern Western cosmology tells us, again, that only up to a certain point was cosmology considered as a scientific way to know the absolute totality of things: during the 18th century, the discourse on totality shifted from an aggregate of sciences (among which celestial mechanics, physics, geometry and mathematics) to another (among which geography, geology, natural

1 Ludwig Wittgenstein, *Tractatus Logicus-Philosophicus* (London: Kegan Paul, Trench, Trubner & Co., 1922), p. 25.

2 See Paolo Virno, *L'idea di mondo. Intelletto pubblico e uso della vita* (Macerata: Quodlibet, 2015).

history and anthropology), putting back the human subject at its center.

It is crucial to notice, as does Paolo Virno,[3] that precisely in this passage did modern Western cosmology disclose its aesthetic nature. Most evidently with Kant, the human access to totality was demoted from the sphere of objective knowledge to that of feelings, and the domain of objective totality shrank to the world that can be actually known and explored. To the question on the "who?", Kant replied by figuring an abstract, a-historical subject capable of *feeling* rather than *knowing* totality *sub specie aeternitatis*. This feeling—which Wittgenstein will later call the "mystical feeling"[4]—is far from aesthetic in a poor and strict sense. It has to do with the harmonic relationship between individual interiority and worldly exteriority, which is indispensable for action. The Kantian feeling of entirety is an anticipation of the community to come and the preparation of the passage to the realm of freedom.[5] Aesthetics, then, is autonomous only *formaliter*: in fact, it prepares practical reason and the affirmation of freedom in the world. We can conclude that when it came into existence, aesthetics was already deeply concerned with the most important problem of its time: how to construct the new world.[6]

The modern passage from a totality that can be *known* to a totality that can be *made* will be treated in detail in the first chapter. For now, I shall emphasize that this attempt at reconciling knowledge and praxis—which Kant simply prepared—goes hand in glove with the concealment of the historical subject of this operation and of its material goals: the universal subject of knowledge built in the centuries before could now take action. For cosmological knowledge to rebecome the knowledge *of* someone, and therefore its power (a refrain that crosses

3 Virno, chapter 2.
4 Wittgenstein, p. 89.
5 See Gianni Carchia, 'Le rovine della rappresentazione. Lettura della «Critica del Giudizio»', in *La legittimazione dell'arte* (Napoli: Guisa, 1982), especially pp. 140–141.
6 I take up Ken-ichi Sasaki's remark in the paper 'The Politics of Beauty' as quoted in Arnold Berleant, *Sensibility and Sense. The Aesthetic Transformation of the Human World* (Exeter: Imprint Academic, 2010), pp. 200–201.

the whole 19th century in a more or less critical sense), it had to dispose of its metaphysical delusions and take on a more markedly ethical and political character. Thus the height of the modern cosmological perversion opened up the possibility of its critique.

Despite the fact that astronomy has restored the scientific meaning of the word 'cosmology' since the beginning of the 20th century, we are now fully conscious that "each epoch develops cultural forms and modes of expression that allow it, however partially and ideologically, to represent its world—to 'totalise' it".[7] I would extend Jameson's remark that "it is ultimately always of the social totality itself that it is a question in representation, and never more so than in the present age",[8] to all regimes of representation, including science. Differently from a strictly Marxist perspective, I tend not to rigidly separate the representations produced through aesthetic technologies from supposedly 'more veritable' representations: although science is evidently less socially determined than aesthetics, the study of cosmology demonstrates that the construction of totality is a mobilization of different apparatuses of experience that concur in delimiting a certain material-symbolical relationship with the world.

Science and aesthetics are placed within a differential continuity of practices and discourses that simultaneously conceal and express the production system of Western society. This is all the more true today, as Jameson himself beautifully acknowledges:

> In a society in which the near-total commodification of the world, linked with the already looming world market, can be glimpsed, beauty is no longer a momentary relief from 'interest' (Kant's word for business and its rationale), but rather the law of the land. For the existential support of universal commodification, even when we try, as so many have done, to grasp it psychologically, was aestheticization always.[9]

---

7 Alberto Toscano and Jeff Kinkle, *Cartographies of the Absolute* (Winchester, UK, and Washington, USA: Zero Books, 2015), p. 14.

8 Fredric Jameson, *The Geopolitical Aesthetics: Cinema and Space in the World System* (Indiana University Press: Bloomington and Indianapolis 1992), p. 4.

9 Fredric Jameson, 'Gherman's Anti-Aesthetic', New Left Review, 97 (2016), 95–105 (p. 95).

The goal of producing and accumulating value is in itself a semiosis, as it transfigures the world based on its own style (an idiosyncratic logic that determines internal tendencies). The base, we could say, is viscous. This is why this book includes aesthetic analyses regarded as fundamental ways of access to the epoch. But aesthetic devices are all the more interesting because they consist in intensifications of the social that may also serve the function of slowing it down and opacifying it. The more an aesthetic device is complex and stratified, the more it diffracts the social structure and offers a chance to counter it—a chance that in itself comes with no explanation, but is sometimes immediately shareable.

This is first and foremost a book about the cosmologies of the present, epitomized by one figure: the planet. Just like the Anthropocene until a few years ago, today the planet is on everyone's lips. "A technoscientific object, a philosophical event", "a call to new forms of governance"[10] and much more, the planet is 'our' form of social totalization. My contention is that the planetary cosmological frame takes its roots in a post-WW2 ideology that relies on the social and epistemic implications of the new technological paradigm of information, and that after the end of the Cold War, when it manifested into a full-fledged cosmological phase, ramified into two lineages: accelerationism and Gaian ecologism. These two positions embody the distinctive features of planetary cosmology and represent its inner contradiction, the irreconcilable contrast at its core. At a logical level, it is a contrast between abstraction and concreteness; at a spatial and political level, between globalism and localism (or rather a hyper-globalist planetarism and a localist planetarism); at a mythological level, between Prometheanism and Epimetheanism; and so on.

The question animating my analysis is also about the cosmology that is needed today: one that does not separate experience from reality while not surreptitiously magnifying the Anthropos; one that provides sufficient reasons to inhabit the Earth without giving up analyzing and confronting the forces that makes it uninhabitable.

10 Nils Gilman, 'The Condition of Planetarity', in Nils Gilman (ed.), *The Planetary* (Los Angeles: Berggruen Press, 2026), 11–34 (p. 11).

However, the present book contains only the *pars destruens*, that is the history and critique of the planetary divarication. The constructive part is not yet there, although it insists like a specter throughout all chapters. So this book leans forward; it sows as it destroys.

1.
COSMOLOGICAL OPERATIONS

1.1 *World (1633)*

The complex mosaic of struggles and aspirations of modernity can also be seen as a nonlinear sequence of cosmological operations. The first one dates back to René Descartes' aborted attempt to construct a Copernican philosophy. In his unpublished treatise on *The World*, written between 1630 and 1633, Descartes declared: "I include the planets and the comets together with the Earth because they, like it, also resist light and reflect its rays, and so I recognize no difference between them".[1] The Earth was proclaimed identical to all other planets, a physical body moving in a space that followed the geometrical laws of Euclid. The totality revealed by Copernicus and Galilei was the physical universe itself, that could now be described in its entirety as a clock-work choreography of shapes, speed and causal relations. From then on, this cosmological framework will be called "the World".

Cosmology is perhaps where Descartes strayed farthest from the mechanical origin of his theories. Whereas the revolutionary advancements made by Copernicus and Galilei had stemmed from very concrete technological problems related to the practice of navigation, Descartes' construction of the World as the homogeneous totality of all things was based on a fundamental caesura from ordinary experience. World cosmology relied precisely on the substitution of natural, qualitative topology (which was at the origin of the Aristotelian distinction between sublunar and celestial world)

1 René Descartes, *The World*, in *The World and Other Writings*, ed. by S. Gaukroger (Cambridge, UK: Cambridge University Press, 2004), p. 20.

with a whole different vision of matter and space based on Euclidean topology. Descartes' intention was indeed to reformulate the Scholastic view of matter through the geometry of particle physics and describe a reality devoid of intrinsic forms, qualities and voids, a purely geometrical and intellectual plenitude that did not resemble the domain of ordinary experience. A universal axiological order was thus replaced by a seemingly neutral plane of repartition. As shown by Châtelet,[2] this collapse of hierarchical distinctions implied not only the exclusion of transcendence in the description of nature, but also the emptying of its potential and vital features.

Taken in itself, Descartes' theory of nature was just simpler and more formally consistent than that of Scholasticism;[3] yet it was also an unprecedented operation of totalization on scientific bases: a monism of matter that eliminated all transcendent forces without including chance. The Cartesian World was a universal and unified totality with no place for contingency and history. This certainly echoed Galilei's dream of total legibility. Unlike Galilei, however, Descartes absolutized the possibility of mathematically abstracting material reality, attempting to realize what the scientist used only as a regulatory ideal. In this way, he translated the materialist revolution of Copernicanism into a rationalist cosmological framework: his cosmological monism depended upon his metaphysical dualism. The tentative discoveries on tides and lunar cycles became the certainties of God's mind and of the knowing soul.

Many words have been spoken on the reasons of this original act of rationalization. The causes of Descartes' rationalist-mechanistic interpretation of Copernicanism surely lie in the emergence of mechanicist ideology in 17th century Europe. The decisive explanatory element, however, is the difference between the rationalist and

2 Gilles Châtelet, *Figuring Space: Philosophy, Mathematics and Physics*, trans. by R. Shore and M. Zagha (Dordrecht: Springer, 2000), chapter 1.

3 Already in *The World*, a dual viewpoint is diplomatically evoked: "Know also that in order to keep my peace with the philosophers, I have no desire to deny that which they imagine to be in bodies in addition to what I have given, such as their *substantial forms*, their *real qualities* and the like; but it seems to me that my explanations ought to be approved all the more because *I shall make them depend on fewer things*" (Descartes, p. 268).

the empiricist approach to science, as embodied for instance by a Frenchman like Descartes and an Englishman like Francis Bacon. Descartes was the expression of a bourgeoisie less in contact with machinery and the problems from which the scientific revolution originally arose, a class still very much tied to the old humanistic values. This is why he amplified the mechanic-mathematical thinking through the old dream of a *mathesis universalis*, which—*pace* Grossman[4]—was a humanist utopia rooted in the bourgeois desire for social stability rather than an emancipatory project for the great masses. The failure of this utopian hope "to unify society as a whole within the design of the development of bourgeois freedom"[5] was symbolized by Descartes' choice not to publish his cosmological treatise after Galileo's sentence in 1633.

While in France the compromise between bourgeoisie and monarchy was giving birth to the machine-state, in England a more dynamic middle class accepted an open confrontation for the reconfiguration of society, which led to the civil war of 1642–51 and the Glorious Revolution of 1688–89. Hobbes, who crafted a political translation of Cartesian cosmology,[6] still expressed the dominant social demand of the first half of the 17th century: peace, order and

---

4 See Henryk Grossman, 'Descartes and the Social Origins of the Mechanistic Concept of the World', in Gideon Freudenthal and Peter MacLaughlin (eds.), *The Social and Economic Roots of the Scientific Revolution* (Springer, 2009), 157–230.

5 Antonio Negri, 'Problemi di storia dello stato moderno', Rivista Critica di Storia della Filosofia, 22/2 (1967), 182–220 (p. 193). Negri argues that Descartes embodied the bourgeois aspiration to translate the civil society into the state after the separation of the former from the latter, a process already begun in the 16th century. In this sense, Descartes' rationalist utopia would be a continuation of the humanist utopia *after* its failure. However, Negri disconnects Descartes from mechanicism and the mechanic-mathematical thinking altogether, through that which seems like an exaggeration of specific aspects of his thought. See also Antonio Negri, *The Political Descartes: Reason, Ideology and the Bourgeois Project*, trans. by M. Mandarini and A. Toscano (London and New York: Verso, 2007).

6 See Andrea Bardin, *Hobbes' Materialist Agenda. The Politics of Early Modern Science* (Edinburgh: Edinburgh University Press, 2026). From an epistemological perspective, the gesture that Hobbes extended to political theory can be summarized as "placing one of the two structural features

the preservation of civil society. A demand of change within the emerging modern state was conveyed instead by the children of the class compromise of 1688,[7] like Locke and Newton. It is only then that Copernicanism gave birth to a successful scientific cosmology as part of a consistent philosophy of nature and a worldview more firmly rooted in empirical knowledge, specific laws (especially the law of gravitation, deducted by Newton from Kepler's third law and the analysis of circular motion by means of centrifugal force) and practical problems (like the irregular movement of chronometers during sea voyages, high and low tides and other problems relevant to navigation). In the third book of Newton's *Principia*, the Cartesian principle of totality (the World) was theoretically maintained, but also anchored to a solid empirical basis. Importantly, empiricist mechanicism reintroduced transcendence in the explanation of nature, because unlike rationalist cosmological monism it did not postulate the unlimitedness of reason.[8]

The importance of the 'loser' Descartes, however, was much greater. His quantitative and speculative cosmology interpreted a movement initiated in the 16th century and culminated in the mid-18th century: a cycle of contraction that prepared industrial capitalism

---

of the new science ahead of the other: the tendency towards deduction rather than towards experimentation" (p. 6).

7 This is an effective paraphrase of Engels by Boris Hessen, 'The Social and Economic Roots of Newton's *Principia*', in Freudenthal and MacLaughlin (eds.), 41–102 (p. 66).

8 The passage from rationalist to empiricist mechanicism, which finds in Hobbes a fascinating connection point, is obviously more complex, and was not at all, as it were, a return to Bacon. Zilsel identifies a specific aspect of continuity: mechanicism already interpreted the separation between observable and unobservable constants as a difference between "real" qualities (e.g., motion) and "apparent" qualities (e.g., color). Locke is a good example of this. "It was the rise of mechanistic physics that turned the harmless distinction between subjective and objective components of observation into a dualism of inner and outer world" (Edgar Zilsel, 'Problems of Empiricism', in George de Santillana and Edgar Zilses, *The Development of Rationalism and Empiricism* [Chicago and London: University of Chicago Press, 1941], p. 69).

after the age of the great explorations.[9] A transition had already taken place to a capitalist "world-economy",[10] which required a new cosmology that universalized the means and values through which this new zone of integration was being built. The rationalist dream of a *mathesis universalis* did not disappear, but migrated elsewhere (for example in the first descriptions of the market as a spontaneous, harmonious and self-regulating order). World cosmology will remain fundamentally torn between rationalist and empiricist tendencies. In fact, no notion of 'totality' would have been allowed by an empiricism consistent with itself; and yet it was precisely the enlarged empiricism of modern science that had shown the impossibility of confining cosmological knowledge to ordinary experience.

Throughout early modernity, the position in the rationalist-empiricist spectrum was determined each time by material, social and historical reasons. In the early 18th-century Prussia, for example—where an unprepared bourgeoisie had to compete with strong power elites and spiritualist movements—, rationalism still played an important role alongside—rather than in opposition to—empiricism, as perfectly illustrated by Christian Wolff's *Cosmologia*

---

9 Wallerstein defines it as a B-phase hegemonic cycle that goes from 1600 to 1750 (see Immanuel Wallerstein, *The Modern World-System II: Mercantilism and the Consolidation of the European World-Economy, 1600–1750* [Berkeley, Los Angeles and London: California University Press, 2011[2]]). Incidentally, it may be noticed that the three cosmological operations I am outlining in this chapter are comparable to the cycles described by Wallerstein: World cosmology from the beginning of the 17th century to the mid-18th century (the hegemonic cycle illustrated in *World-System II*), Earth cosmology from the 18th century to WW1 (the geocultural period of Wallerstein's *World-System IV*, embracing several Kondratieff cycles), System cosmology from WW2 until the 1970s (within the fourth Kondratieff cycle).

10 "A large geographic zone within which there is a division of labor and hence significant internal exchange of basic or essential goods as well as flows of capital and labor" that is "not bounded by a unitary political structure" (Immanuel Wallerstein, *World-System Analysis: An Introduction* [Durham and London: Duke University Press, 2004], p. 23). The zone of world-economy is not necessarily (and not yet) the entire globe: it is a relatively isolated zone of integration of different political structures and cultures.

generalis (1737). In England, empiricism had simply proved itself a better ideology for the social compromise between middle class and power elites, as the battle had to be fought on the ground of science and fungible reason rather than abstract knowledge.

A functioning reunion of the rationalist and empiricist strands of cosmological mechanicism will occur only a century later. Let us take Pierre-Simon Laplace's illustration of the idea of a deterministic universe. A political opportunist and incapable administrator, Laplace's work reflected a decisive change: as the French bourgeoisie truly seized power in Europe, cosmology became an instrument of government. In his *Exposition of the World System*, partly written during a retreat in the countryside amid the events of the 1789 Revolution, Laplace argued that the ideal intelligence which can determine the entire evolution of the universe from the knowledge of one instantaneous state cannot be attained by human intelligence except through a "feeble likeness", but can indeed be harnessed through the calculus of probabilities. The calculus of probabilities was presented as a form of mastery of the universe which determines the superiority of humans over animals and the progress of nations.[11] Finally, the Cartesian World could become an instrument for those who took the risk of enterprise according to a rational calculus.

World cosmology, however, was an all too old instrument for the new ruling class. Newton had already introduced the cosmology of the modern subject, the I that navigates reality in practice rather than in the abstract. It is in this same spirit that Kant, in his first *Critique* (1781), reduced the concept of World to an antinomy of reason and prohibited to include it in reasonable discourses. The metaphysical use of the term "World" with reference to the absolute totality of things by philosopher such as Leibniz, Wolff and Maupertuis was ultimately acknowledged as illegitimate. Kant's cosmological prohibition was, after all, the natural outcome of a philosophy which desired to bring rationalism and empiricism to terms within Copernicanism and Newtonianism.

---

11 Pierre-Simon Laplace, 'On Probability', in *Philosophical Essays on Probabilities*, trans. by A.I. Dale (New York: Springer, 1995), pp. 2–3.

The abolition of the World was the culmination of a shift in cosmological sensibility that had been at work for the whole 18th century. Particularly symptomatic in this sense was the gradual replacement of the concept of "World" with that of "world(s)", indicating an astronomical multiplicity of world-bodies (*Weltkörper*), still seen mostly as perfect spheres. In 1771, Swiss polymath and astronomer Johann Heinrich Lambert thus summarized the state of the cosmological discourse: "We find worlds on every speck of dust, every drop, and soon the specks of dust will be barely so numerous as the terrestrial globes in the firmament".[12] As early as 1734, the possibility that "other worlds are born according to the same causes, modalities, contingencies and failures that brought our own planet to life"[13] was already relatively diffused. Celestial mechanics had ended up breathing new life into the ancient idea of the plurality of worlds and this because the universe was starting to be seen not through the lens of mathematics, but of human life and history. It was now conceivable—even probable—that other planets were indeed Earths[14] (an idea that Descartes explicitly negated in the *Principles*

12 Johann Heinrich Lambert, *Cosmologische Briefe über die Einrichtung des Weltbaues* (Augsburg: Eberhard Kletts Wittib., 1771), p. 108.

13 Emanuel Swedenborg, *Principia Rerum Naturalium sive Novorum Tentaminum phenomena mundi elementaris philosophice explicanda*, in Id., *Opera Philosophica et Mineralia*, Vol. I (Dresden and Leipzig: Friedrich Heckel, 1734), p. 24.

14 "We see that all the planets", wrote de Fontenelle in 1686 fantasizing on the Copernican worldview, "are of the same nature, all opaque bodies that receive light only from the Sun and reflect it from one to the other, and have nothing but the same motions; up to that point, everything is equal. Yet we are expected to believe that these great bodies should have been fashioned not to be inhabited, that this should be their natural condition, and that there should be an exception made in favor of the Earth alone" (Bernard le Bovier de Fontenelle, *Conversations on the Plurality of Worlds*, trans. by H.A. Hargreaves [Berkeley, Los Angels and Oxford: University of California Press, 1990], p. 44). It must be recalled that in itself, the idea that every planet was inhabited is far more ancient and was present, in a Copernican sense, since Copernicus himself.

of Philosophy). With the improvement of telescope technology, the universe now resembled a "luxuriant garden".[15]

The shift from the infinite World to the many singular worlds had one meaning: all universe started to be seen from the viewpoint of the Earth, regarded as 'the world'. This was perfectly compatible with Copernicanism, as the Copernican revolution did not begin by stating that the Earth was identical to the other celestial bodies, but conversely, that the celestial bodies were similar to the Earth (i.e., corruptible rather than heavenly). As a consequence, all planets became scientifically comparable to the Earth. In reaction to the decentering abstraction operated by Descartes and in order to mend the opposition between the totality accessible to the soul and the restricted domain accessible through the senses, it was now time to rebuild a unitary cosmology around the I. The World was no longer suitable for a natural order characterized by relentless change.[16] It was the synthesis brought about by the European bourgeoisie, now definitively established as the new ruling class.

1.2. *Earth (1934)*

The World, maintained Kant, is "the mathematical whole of all appearances and the totality of their synthesis".[17] The world, instead, corresponds to the human being and its environment, the Earth. The interior and the exterior domains were now part of the

---

15 Frederick William Herschel, quoted in Michael Hoskin, *William Herschel and the Construction of the Heavens* (London: Oldbourne, 1963), p. 115.

16 As Clark notes in the wake of Collingwood, the cosmology of the World "by necessity excluded change as part of the natural order. Change was omitted by either abstracting changeable aspects of phenomena or by concentrating on permanent unchanging relations. The permanency of change by the eighteenth century thus required a new cosmology" (Charles M.A. Clark, *Economic Theory and Natural Philosophy. The Search for the Natural Laws of Economy* [Aldershot and Brookfield: Edward Elgar, 1992], p. 24).

17 Immanuel Kant, *Critique of Pure Reason*, trans. and ed. by P. Guyer and A. W. Wood (Cambridge, UK: Cambridge University Press, 1998), pp. 465–466.

same cosmology.[18] The world is the Earth as the individual's space of action and experience: "The substrate and the theatre in which the game of our ability is played" and "the ground on which our knowledge is acquired and applied".[19] It is the interiorization of the Cartesian universe after the empiricist correction.

By placing human interiority at the center of the system of nature, Kant re-included the dimension of time, which had no place in the Cartesian universe. Already in Newtonian cosmology, the physical existence of the void and a growing awareness of earthly geology sustained the idea that nature was crossed by forces of decay and that (contrary to what Leibniz will later object) the universe was not a great machine. This was a fundamental step towards the affirmation of human freedom in a world governed by physical laws *and* by God. As notes Virno, "the thinking of the world presupposes the thinking of the Subject and prepares the thinking of God".[20]

The passage from the World as the universal totality of objective reality to the world as the singular totality of subjective reality prepared the epochal cosmological shift already fully visible in Kant: the re-centering of the Earth. The world *is* the Earth: a singular, dynamic totality that grounds the human I. The re-moralization of reality was a consequence of the immanentization of Copernican totality. The discourse on totality was now attributed to another aggregate of sciences guided by anthropology and physical geography, and astronomy was no longer the leading knowledge of cosmology.

Earth cosmology will be perfectioned after Kant by rationalist spiritualists such as Hegel and Schelling and by empiricist

---

18 The world (*Welt*) as the subject of physical geography is thus described: "The world as *object of external sense* is nature, as *object of internal sense* is the soul or the human being" (Immanuel Kant, *Physische Geographie*, Erster Band, in *Kants Gesammelte Schriften*, Erste Abt., Band IX, hrsg. von der königlich Preußische Akademie der Wissenschaften [Berlin and Leipzig: De Gruyter, 1923], p. 156, my translation). Note that in German, the initial of *Welt* is always capitalized.

19 Kant, p. 158 (my translation). It is worth remembering that after an interest in cosmology during his youth, when he formulated the famous nebular hypothesis later refined by Laplace, Kant taught physical geography for forty years.

20 Virno, p. 33.

spiritualists like the German *Naturphilosophen*. Hegel struck at the very heart of Cartesian cosmology by arguing that the Earth is not at all a celestial body among others, but an individuated *potentia* (in German: "*Macht*")[21] that evolves into human history and universal reason. In the same vein, Schelling de-somatized the planet by endowing it with the ontogenetic potency that belongs to nature in general. Hegel's and Schelling's use of the term "world", as in the concepts of *Weltgeist* and *Weltseele*, was evidently centered on the life of the Earth and the history of nature as it is expressed on Earth. The Earth was now a 'body that moves' in a different sense: not a physical body endowed with motion, but a speculative organism endowed with Aristotelian motility.[22]

The spiritualist conception of the Earth was nourished by the spreading of a historical view of nature within sciences and by the bustling social scenario of the time. When Hegel wrote his philosophy of nature, Germany was torn between liberal demands and absolutist reaction; the new substance of the social body, however, had already manifested itself: neither State nor multitude, but the pure movement of the bourgeois class. Yet, this movement still needed to exist besides the previous political élites (especially in a rapidly growing but still pre-industrial nation like the German Confederation). With this social configuration in his heart and mind, Hegel was irritated and disconcerted by the 1830 insurrections, which negated the delicate balance achieved. Particularly telling in this sense is his use of the

---

21 "The Element of developed difference and its individual determination is Earth: as distinct from the other moments, it is in the first instance earthiness as such, and as yet indeterminate; but, as the totality which holds together the different moments in individual unity, it is the power [*Macht*] which kindles them into a process which it also sustains" (Georg Wilhelm Friedrich Hegel, *Philosophy of Nature: Part Two of Encyclopedia of the Philosophical Sciences (1830)*, trans. by A.V. Miller [Oxford: Clarendon Press 2004], p. 113).

22 In Hegel's *Encyclopedia*, mechanical motion is famously associated with lack of freedom. And yet, celestial mechanics is regarded by Hegel as a domain of absolute freedom and the motion of the planets as "*absolutely free*" (Hegel, p. 61). This is because planets are considered as abstract individualities in a speculative sense. See Antonio Moretto, 'Negazione e libertà nella teoria hegeliana del meccanismo', Verifiche, XXVIII (1999), 3–23.

word "*Macht*" to describe the Earth's *potentia*: the German "*Macht*" reunites *potentia* (the force to produce and act) and *potestas* (the faculty and right to produce and act). Hegel's Subject is a force that immediately becomes right: not a transcendent authority nor a telluric drive, but that immanent force of structuration which corresponds to the life of the modern bourgeois society as described in the *Elements of the Philosophy of Right*. The static, metaphysical *mathesis* dreamt by Descartes—symptom of an earlier and more conflicted stage of the capitalist world-system—was long gone.

For the first time, the Earth is explicitly characterized as "the *abstract* ground of individuality".[23] The cosmos definitively lost its character of physical totality and gained the status of a spiritual ground, more temporal than spatial, crossed by the abstract potency that progressively individuates itself into experience and thought. Criticizing Descartes and Newton, Hegel explicitly claimed that rationalist cosmology and celestial mechanics had made the universe incommensurable to subjective—i.e., individual—experience.[24] True rationality, he argued, deepens and elevates experience rather than alienating it. It was therefore necessary to rebuild cosmology on the basis of its ontological (and therefore also epistemological) continuity with subjective individuality. Hegel's attempt to reconcile rationalism and empiricism was a different form of rationalism, a different vision of total fungibility—not physicalist, but spiritualist: the reduction of all things to the non-identical as the total superimposition of value.

This conceptual shift reflected the vectors of constitution of a new scientific paradigm, itself dense with cosmological implications: thermodynamics. When Clausius stated in 1865 that "the energy of the World is constant" and "its entropy tends to a maximum", he employed the concept of "*Welt*" with the old meaning of universal totality, but in a completely different sense. The universe was no longer infinitely open, static and geometric, but closed, energetic and dynamic—an elemental cosmos once again after the Copernican revolution. The thermodynamic universe partly continued Copernicus' work of

23 Hegel, p. 123.
24 Hegel, §286.

demystification, thus going beyond the spiritualist cosmologies: the universal *dynamis* was not a speculative principle of life, but one of material consumption. The fact, already noticed by Newton, that the universe is crossed by degradation tendencies was elevated into a full-fledged cosmological principle. The universe acquired a history marked by the incessant transformation of matter into energy: a history directed not towards eternity, but towards cosmic death. Whereas in the *Naturphilosophie* the heat represented a vitalistic universal principle and a dynamic substance (exemplarily in Franz von Baader's theory of *Wärmestoff*), now heat was a principle of loss. The universe was not only corruptible; it was also irreversible.

On the other hand, thermodynamics further developed the idea that the Earth is the abstract ground and stage of human individuality—or, in philosophical terms, an abstract individuality.[25] In Clausius' decaying universe, the Earth is indeed a singular exception: a planet-size machine, a miracle of order within a chaotic reality. It is no chance that the discoverers of thermodynamics deduced their cosmological theories from the Earth regarded as an ensemble of energy-producing systems, and that later on, thermodynamics will be much more useful to describe the Earth than the universe. The thermodynamic paradigm broke in two the principle of equivalence introduced by Copernican cosmology: the heavens are indeed like the Earth (they are made of the same matter), but the Earth is not like

25 This conceptual passage, clearly entailed by Hegel's theory of the Earth and by the Idealist philosophy of nature, is even more clearly formulated in the first pages of Feuerbach's *The Essence of Christianity* (1841), where the Earth is regarded in relation to "its own nature" according to a subject-object relationship (which, however, Feuerbach extends to all planets, in a reminiscence of world cosmology). "The relation of the Sun to the Earth is therefore at the same time a relation of the Earth to itself or to its own nature, for the measure of the size and of the intensity of light which the Sun possesses as the object of the Earth is the measure of the distance which determines the peculiar nature of the Earth. Hence each planet has in its sun the mirror of its own nature" (Ludwig Feuerbach, *The Essence of Christianity*, trans. by M. Evans [London: Trübner & Co., 1881[2]], p. 5). Here Feuerbach takes Hegel's Earth theory almost literally ("The process of the earth is perpetually kindled and stimulated by its *universal self*, the activity of light…", Hegel, p. 118).

the heavens (due to its history *as* matter). Insofar as it demonstrated that all things irremediably decay, this new worldview imparted an exemplary anti-Idealist lesson and fueled the most profound nihilism in the European bourgeois societies *fin-de-siècle*, anticipating the end of their appropriated wealth. But insofar as it demonstrated the miraculous uniqueness of the Earth as an anti-entropic oasis, it prolonged the Idealist perspective and set the bases for later spiritualist reappropriations.

That said, thermodynamics has also been the cosmology of the proletarian class, and the adversarial continuity between Hegel and Clausius was also that between Hegel and Marx. Before the Earth as an ensemble of energy-producing systems, the origin of thermodynamic cosmology was the mechanics of industrial machinery; the first subjects of thermodynamics were theoretical machines, like the Carnot machine. Just like Copernican cosmology was rooted in the mechanical problems of navigation during pre-industrial capitalism, thermodynamic cosmology was rooted in the mechanical problems of industry. While the former was the practical cosmology of the merchant (and was later abstracted based on the ideological model of manufacture), the latter was the practical cosmology of the worker. For this reason, Engels, Lenin and Stalin integrally adopted the thermodynamic cosmology (with the only exceptions of Clausius' extension of the second law to the whole universe, that is of the concept of heat death, and the idea of the universe's finiteness).

The Marxist-Leninist universe was infinite and eternal, and with the Earth at its center. As wrote Engels,

> the geocentric standpoint in astronomy is prejudiced and has rightly been abolished. But as we go deeper in our investigations, it comes more and more into its own... The sun, etc., serve the earth, Hegel, Philosophy of Nature [...]. Anything other than geocentric physics, chemistry, biology, meteorology, etc., is impossible for us, and it loses nothing by the phrase that this only holds good for the earth and is therefore only relative. If one takes this seriously and demands a

> centreless science, one puts a stop to all science; [it suffices] us to know that under the same conditions everywhere the same (...)![26]

In the exact same vein, Marx urged to abandon the cosmological question once it becomes "impossible in practice" and reformulate it in terms of the human power of self-transformation through the active and sensuous relationship with the world. For both Marx and Engels, the thermodynamic universe is the historical universe, at the center of which lies the historical Earth. This is why geology is so important in demonstrating that our planet, like all planets, has a genesis and developed through a "process" of "self-generation" rather than being created by God.[27] However, these facts must not be turned into new absolutes. Once established that humankind has a historical essence, the speculative question must be abandoned. "When you ask about the creation of nature and man, you are abstracting, in so doing, from man and nature. [...] Now I say to you: Give up your abstraction and you will also give up your question".[28]

It is here, with Marx and Engel's interpretation, that the second cosmology of modernity developed a response to its original spiritualist character and its inner contradiction. Like World cosmology was contested between rationalism and empiricism, Earth cosmology—the cosmology of the modern I—was contested between spiritualism and materialism. Both these positions recentered the universe on the Earth and humanity; but while the spiritualists saw them as a metaphysical center and pinnacle of speculative life, the materialists took them as the only relevant starting point for praxis.

---

26 Frederick Engels, *Dialectics of Nature*, trans. and ed. by C. Dutt (New York: International Publishers, 1940), p. 260.

27 Karl Marx, *Economic and Philosophic Manuscripts of 1844 and the Communist Manifesto*, trans. by M. Milligan [New York: Prometheus Book, 1988], p. 112). Science tells us about "nature's coming to be man" (p. 111). Engels attributes the discovery of geogony and cosmogony to Kant in the *Universal Natural History and Theory of the Heavens* from 1755: "The earth and the whole solar system appeared as something that had come into being in the course of time" (Engels, pp. 8–9).

28 Marx, p. 113. In the first pages of the *Economic and Philosophic Manuscripts*, the word "earth" appears already after the abstract question has been abandoned as another name for land before appropriation.

The highest moment of clarity of Earth cosmology came in a spiritualist context, at the climax of the European crisis, when this cosmological operation was already shifting away into something else. In an unpublished writing from 1934, Edmund Husserl moved a critique to the Cartesian universe that closely resembled Hegel's: the Earth cannot be abstracted into a physical equivalence with the other planets, because unlike other planets it is the source of *our* experience. By negating our experiential belonging to the Earth and posing a fictitious universal viewpoint, the Copernican paradigm has broken the link with lived experience. But science should account for the fact that the Earth is the constitution of the transcendental I itself in its historical making, our "primordial home",[29] which is also our history. The Earth, argued Husserl, is humanity's "originary ark": an elemental Nature immobile and mobile at the same time—that is, a transcendental foundation.

Husserl still made use of the weak notion of "world" in the Kantian sense; he even admitted the idea of other Earths, although as related to the original Earth, since "there is only one humanity and one Earth".[30] But by explicitly countering Copernicanism and rehabilitating the "Ptolemaic folly", he asseverated the philosophical core of Earth cosmology and set the terms of its overcoming. Confronted with the thermodynamic truth that the Earth itself, like all physical entities, will eventually die, he retreated in a consolatory notion of humanity: the Earth may die, he wrote, but it will also continue to live as long as its history continues, just like my individual body and its "We".[31] This tautological avoidance of the thermodynamic

29 Edmund Husserl, 'Foundational Investigations of the Phenomenological Origins of the Spatiality of Nature: The Originary Ark, the Earth, Does Not Move', in M. Merleau-Ponty, *Husserl at the Limits of Phenomenology. Including Texts by Edmund Husserl*, ed. by L. Lawlor and B. Bergo (Evanston: Northwestern University Press, 2002), p. 126.

30 Husserl, pp. 127–130.

31 "It is possible that entropy will put an end to all life on earth, or that celestial bodies will crash into the earth, etc. But even if one found in our attempts the most unbelievable philosophical hubris, we would not back down from the consequences for the clarification of necessities pertaining to all sense donation for what exists and for the world. We do not back down even when confronting the problems of death in the new way

evidence reveals the moral drop point of phenomenology as part of the post-Copernican re-moralization of reality, now in the guise of a spiritualist humanism stemming from an original combination of Kant and Hegel.

This philosophical recipe will be perfectioned by Husserl's most brilliant disciples, Martin Heidegger and Maurice Merleau-Ponty. Heidegger developed Husserl's conception of Earth as Nature after employing the concept of 'world' in the Kantian sense in *Being and Time*. Heidegger's Earth is a *physis*, that is nature taken in its elemental and genetic power of expression, inseparable from the world as its actual expression. Echoing Hegel's critique of Descartes, his Earth cosmology revolved around a notion of *potentia* taken from Aristotle's concept of *hyle*.[32] Heidegger's appeal to nature was a way to draw new forces beyond the liberal civil society, far from the social compromises envisioned by Hegel and Husserl himself.

Drawing explicitly from Husserl's writings from a more liberal perspective, Merleau-Ponty later came to similar conclusions. In a note from 1960 to his unfinished work *The Visible and the Invisible*, he hinted at the project of a "transcendental geology",[33] which would

---

phenomenology conceives them. In the present, I as something present am progressively dying, others die for me when I do not find a present connection with them. But a unity by recollection permeates my life—I still live, although in being other, and continue to live the life that lies behind me and where its sense of being behind me lies in reiteration and the ability to reiterate. Thus the We lives in the reiterability and continually lives in the form of reiterability of history while the individual 'dies,' that is, the individual can no longer be 'remembered' empathically by others, but only by the historical memory in which the subjects of the memory can be represented themselves" (Husserl, p. 131). The full consequences of this vague passage on the survival of memory after the Earth's (and the Sun's) collapse will be drawn only much later by François Lyotard in 'Can Thought Go on Without a Body?', in *The Inhuman: Reflections on Time*, trans. by G. Bennington and R. Bowlby (Stanford: Stanford University Press, 1991).

32 See Martin Heidegger, *On the Essence and Concept of φύσις in Aristotle's* Physics *B, I*, trans. by T. Sheehan, in *Pathmarks* (Cambridge: Cambridge University Press 2013), pp. 186ff.

33 Maurice Merleau-Ponty, *The Visible and the Invisible*, ed. by C. Lefort, trans. by A. Lingis (Evanston: Northwestern University Press, 1968), p.

be something like a "psychoanalysis of Nature",[34] an investigation of the Earth's dynamic ontology of manifestation. The Earth is not a thing with an inside and outside, he argued, but a process that unfolds by folding onto itself. Humankind is that interior exteriority that continues the Earth as Nature's process, its last and most sensitive surface, ultimately destined to fold on itself as well. Like Heidegger, Merleau-Ponty put philosophy in charge to study the Earth as "something that is realized through man" but does not end in anthropology.

In the wake of Husserl, Heidegger and Merleau-Ponty completed the historicization of Hegel and Kant's systems: Hegel's dynamic ontology was devoid of teleologism and brought back to a process of pure historical expression; the Kantian transcendental subject was no longer an abstract structure of faculties but a concretely existing Earth-dweller. Still, the Earth remained a moral center of ontological intensification in an otherwise meaningless universe, and humanity the force that actualizes the history of the universe. The reversibility between interiority and exteriority thematized by Merleau-Ponty was indeed mobilized around a linear axis as a spiral concrescence of form and matter, whose orientation coincides with its manifestation: the evolution of Earth into the Anthropos.[35] Of course, both Heidegger and Merleau-Ponty proved to be aware that the human and its present epoch was not to be magnified and metaphysically justified (at the risk of slipping back into teleology). Heidegger in particular was notoriously suspicious of the human instruments and institutions of his time. However, they both identified the only chance for grace in a spiritualist naturalism turned towards an immemorial future-past: history as eternity, nature as history—as long as "We" exist. "It is the sphere of the spiritual. All social forms, all concretizations, all fixations are foreign to it. Those forms are too noisy, too rigid, too

259.

34 Merleau-Ponty, *The Visible and the Invisible*, p. 267.

35 Merleau-Ponty described it as an "ascent on the spot" moving from the inorganic to "the 'psycho-physical'", a "stabilized explosion" canalized around a temporal axis (Merleau-Ponty, *The Visible and the Invisible*, pp. 177, 268).

palpable for it. It is only in reflection, in solitude, in silence above all, that we access it. But we aspire to it in all our being".[36]

1.3. *System (1979)*

Earth cosmology was less a "Ptolemaic counter-revolution"[37] as it was one possible development of the Copernican revolution and Descartes' dualism. Yet its spiritualist abstraction, crucial to the establishment of centrist liberalism as the ideological framework of modern geoculture,[38] still eschewed the question of finitude, which thermodynamics brought back violently to the fore. Although participating in the fundamental direction of Earth cosmology, thermodynamics introduced a materialist thorn in the side of its main ideological regime that triggered a further de-phasing in Western cosmology. An astronomical description of the universe matching the cosmological materialism of thermodynamics will emerge only decades later, with Einstein's relativistic cosmology (first introduced in the 1917 *Cosmological Considerations*) and the observations made by Hubble during the 1920s.

The universe of Einsteinian cosmology is finite (like a thermodynamic isolated system), but also unlimited; it has an oriented history in a cosmogonic sense, and as such it cannot be pre-determined. It no longer 'contains' bodies, as in Newtonian celestial mechanics, but is in itself determined by the mass of the bodies, which are now to be conceived of as lines of becoming ("worldlines"). The matter of the universe becomes indistinguishable from the forces it exerts and the energy it conveys: matter and energy are two

36 Eugène Minkowski, *Vers une cosmologie. Fragments philosophiques* (Paris: Payot, 1999), p. 258.
37 See Quentin Meillassoux, *After Finitude: An Essay on the Necessity of Contingency*, trans. by R. Brassier (London and New York: Continuum, 2008), chapter 5.
38 See Immanuel Wallerstein, *The Modern World-System IV: Centrist Liberalism Triumphant, 1789–1914* (Berkeley, Los Angeles and London: University of California Press, 2011).

interchangeable states of the same "cosmic fluid" of matter-energy[39] within which world-paths emerge, whose distribution determines the metrical properties of the related region of spacetime.

Another fundamental cosmological shift can be noticed along this passage. On one hand, the Earth was an exception to physical laws and could therefore be seen as the moral center of the universe; on the other, thermodynamics confirmed the anti-anthropomorphic lesson of Copernicus: reality is at its core dissimilar from ordinary human experience and imagination. Thus, a new chasm was opened between human experience and physical reality, exacerbated by the Einsteinian application of a non-intuitive geometry of space (Riemannian rather than Euclidean) and a non-psychological understanding of time to the description of the universe. The fundamental function of commensuration represented by the cosmological operation of the Earth was lost. The idea of a physical cosmos that can be inhabited began to move away from the Western conscience—hence the many paeans on our "worldlessness" that populate the phenomenological literature.[40]

---

39 Jacques Merleau-Ponty and Bruno Morando, *The Rebirth of Cosmology* (New York: Knopf, 1976), pp. 198–199.

40 "There is no longer any world", lamented Jean-Luc Nancy, "no longer a *mundus*, a *cosmos*, a composed and complete order (from) within which one might find a place, a dwelling, and the element of an orientation" (Jean-Luc Nancy, *The Sense of the World*, trans. by Jeffrey S. Librett [Minneapolis and London: University of Minnesota Press, 1997], p. 4). A few years later, Derrida analyzed the question of worldlessness in *The Beast and the Sovereign*, coming to the conclusion that what was lost is "*the* world, *a* world, a world that is *one*" (Jacques Derrida, *The Beast and the Sovereign*, vol. 2, ed. by M. Lisse, M. L. Mallet and G. Michaud, trans. by G. Bennington [Chicago and London: The University of Chicago Press, 2011], p. 367). Here further historical vectors intervene; but the roots of this cosmological loss lie in the development of the cosmological dialectic of World and Earth. The widespread belief of the scientific and epistemological nature of this development is expressed by the ironic words of Lacan thirty years earlier: "The world is in decomposition, thank God. [...] Even in scientific discourse, it is clear that there isn't the slightest world. As soon as you can add something called a 'quark' to atoms and have that become the true thread of scientific discourse, you must realize that we are dealing with something other than a world"

Before the end of the 19th century, the dephasing triggered by thermodynamic materialism already produced compromise solutions that looked beyond the spiritualism/materialism divide. Russian Cosmism, for example, introduced a thermodynamic religion that understood the universe as fully fungible in a material—rather than intellectual—sense. If the Earth is an engine inhabited by engines, as Fëdorov proclaimed and if human work and existence are functionally the same as Nature itself, incessantly transforming matter into energy and energy back into matter, then humanity can materially reach and exhaust all reality. As a consequence, the human being is destined to colonize space (the extraterrestrial universe) and time (eternity through resurrection). "The Earth is the cradle of humanity, but one cannot live in the cradle forever".[41] Not dissimlarly, Comte's project for a positive polity imagined the Earth as the "*Grand Fétiche*" of a new humanist and materialist religion, that is as the objectual mediation between humanity and its cosmic environment (or as cosmic machine of humankind). Husserl's "originary ark" is the archetype of this cosmological juncture: the Earth as a spiritual machine, the spaceship of history.

But the tendencies towards overcoming the second cosmology will materialize in full only after World War 2, when thermodynamics regained its technological nature and evolved into a theory of information and self-regulating machines, soon extended to the study of self-regulating systems in general. Regarded as a hyper-system, the Earth could now be described as continuous to the universe according to its first order properties (that is, for instance, as made of matter-energy), but also as discontinuous to the universe according to its second order properties, its systemic features (mainly its capacity to maintain its own equilibrium on a physical, chemical and biological level, a feature which had already been observed by 19th-century thermodynamics). The two paths opened by thermodynamics had been both undertaken. The Earth was now

---

(Jacques Lacan, *The Seminar: Book XX. Encore 1972–1973*, ed. by J. A. Miller, trans. by B. Fink [New York and London: W.W. Norton & Co., 1998], p. 36).

41 Konstantin Ciolkovskij, *Циолковский К. Э. Избранные труды* (Moscow: Academy of Sciences of the Soviet Union, 1962), p. 207.

a self-organizing region of matter-energy, a particular worldline or self-regulating behavior: a planetary system.

At a social-historical level, all this was made possible by the renewed alliance between State and bourgeoisie that took place in the US during the Cold War. Faced with unprecedented problems of internal and external control, the US refined the wealth of European technoscientific knowledge to develop a system of infrastructural governance that could restructure the economic and social system on the new scale of conflicts and production. Systemic thinking was indeed a way to elevate Fordism into a general description of reality (a 'science of sciences') through a new concept of control which corrected Romantic organicism with thermodynamics.

In 1979, after a decade of work with Lynn Margulis, James Lovelock published *Gaia: A New Look at Life on Earth.* This book introduced the public to the so-called Gaia hypothesis, in many ways the symbol of systemic cosmology. The neo-cosmic figure of Gaia was based on a description of the Earth as a complex hyper-system capable of maintaining itself far from equilibrium by regulating the processes of energy transformation that couple it with the cosmic environment. In Lovelock's vision, the Earth system is tendentially homeostatic: a *machine à gouverner*. The analogy with subjectivity was broken: the planet's agency here is not the natural abstraction of morality, but a simple material fluctuation.[42] Yet, there was no relapse into pure matter, not even in the sense of the matter-energy of the new physics, as the least significant entity was the systemic whole, or the systemic complex behavior, a relational and dynamic determination that populates every level of reality. The new cosmological transcendental took the appearance of a "*plissé fractale*",[43] a matryoshka of universe-niches.

42 See Ilya Prigogine and Isabelle Stengers, *Order Out of Chaos: Man's New Dialogue with Nature* (London: Verso, 2018), chapter 7.

43 Pierre Lévy, 'Plissé fractal ou comment les machines de Guattari peuvent nous aider á penser le transcendental aujourd'hui', Chiméres, 8/21 (1994), 167–180. Writes Lovelock: "Whether we are considering a simple electric oven, a chain of retail shops monitored by a computer, a sleeping cat, an ecosystem, or Gaia herself, so long as we are considering something which is adaptive, capable of harvesting information and of storing

The shift to this third cosmology carried with it immediately visible contradictions. The same framework designed to harness instability had to admit instability as a possibility of the system. Thus, the fluctuating planetary behavior was inherently contested between vectors of identification and vectors of alteration. The dialectic of these vectors marked the cultural history of System cosmology. Within the limits of our historical overview, we can distinguish between System cosmologies that cultivate the idea that planet Earth is indeed a system and should be treated as such—let's call them 'pro-system cosmologies'—and System cosmologies that polemize with this idea—or 'anti-system cosmologies'. This brings us to the threshold of the present.

Pro-system cosmologies are cosmologies which rely on Earth system science and Earth system governance without interrogating—and possibly repurposing—their social and epistemic premises. These positions nurture the idea that the Earth can and should be controlled and governed according to its systemic description. Some of the most immediately evident distortions of this belief are gathered under the idea of "geoengineering", whose "fundamental fantasy claims that the Earth, and everything contained on it—ecosystems and organisms, humans and nonhumans—can and must be reconstructed and entirely remade".[44] The hyper-modernist character of ecological constructivism resounds in large-scale governance projects that promise to solve the climate crisis through ecopoiesis, or the direct modification of earthly environments. In this sense, System cosmology inspires a great denial of finitude and enables new forms of trans-scalar reductionism. Ecopoiesis (the belief that environments can be engineered) is a direct consequence of the ideological expansion of the notion of autopoiesis (the description of systems as functionally closed and of their behavior as aimed at self-stabilization and self-regulation). As Lovelock often repeats, the cybernetician's viewpoint is the engineer's viewpoint.

---

experience and knowledge, then its study is a matter of cybernetics and what is studied can be called a 'system'" (James Lovelock, *Gaia: A New Look at Life on Earth* [Oxford: Oxford University Press, 2000], p. 57)

44 Frédéric Neyrat, *The Unconstructable Earth: An Ecology of Separation* (New York: Fordham University Press, 2019), p. 2.

Anti-system cosmologies are based on the disbelief in the normative use of descriptive categories such as homeostasis and self-regulation. A certain kind of anti-system cosmologies criticizes the systemic objectification of the Earth by claiming that "Gaia cannot be compared to a machine".[45] This cosmological position, represented by the likes of Bruno Latour, Isabelle Stengers and Dipesh Chakrabarty, aims at re-centering the universe onto the planet regarded as an infinitely heterogeneous domain, "a gigantic discordant harmony, mutable and contingent".[46] Haraway's sympoietic Gaia, Stengers' "intrusive" Gaia, Latour's "outlaw" Gaia, the "pluriversal" Earth are all figures of a manifold planet that has been recentered on itself.[47] "The sublunary becomes once more distinct from the superlunary, and the idea of world recovers a radically closed sense, which is also to say that it becomes immanent: terrestrial, local, proximate, secular, nonunified".[48] Basing on spiritualist premises regarding the irreducibility of existence and the moral priority of dwelling and conveying the widespread intuition that the univocal world of globalization would be replaced by an infinity of "untranslatable worlds",[49] these authors put forward a kind of a-historical materialism that has gained traction over the past decade.

---

45 Bruno Latour, *Facing Gaia: Eight Lectures on the New Climatic Regime*, trans. by C. Porter (Cambridge and Malden: Polity Press, 2017), pp. 96–97.

46 Eduardo Viveiros de Castro and Deborah Danowski, 'Humans and Terrans in the Gaia War', in Marisol de la Cadena, Mario Blaser (eds.), *A World of Many Worlds* (Durham: Duke University Press, 2018), 172–203 (p. 180).

47 See respectively Donna Haraway, *Staying with the Trouble: Making Kin in the Chthulucene* (Durham: Duke University Press, 2016); Isabelle Stengers, *In Catastrophic Times: Resisting the Coming Barbarism* (Lüneburg: Meson Press, in collaboration with Open Humanities Press, 2015); Latour, *Facing Gaia*; and Marisol de la Cadena and Mario Blaser (eds.), *A World of Many Worlds*. On the notion of pluriverse, see e.g. Ashish Kothari, Ariel Salleh, Arturo Escobar, Federico Demaria, Alfredo Acosta (eds.), *Pluriverse: A Post-Development Dictionary* (New Delhi: Tulika Books, 2019).

48 De Castro and Danowski, p. 179.

49 Jacques Derrida, 'Fidélité à plus d'un. Mériter d'hériter où la généalogie fait défaut', Cahiers Intersignes 13 (1998), p. 347.

On the other hand, authors who define themselves as "accelerationists" and "speculative realists" aim at further de-centering the Earth into the cosmos. The accelerationist "is concerned not with a true-to-earth thought but a thought whose topos is a realist true-to-the-universe earth".[50] This sort of depoliticized cosmism polemizes with the modern paradigm of the Earth, rightly seen as the cosmological projection of the transcendental I, by arguing that the planet belongs to the radical alterity of extraterrestrial space. Authors like Ray Brassier, Eugene Thacker and Ben Woodard consider the planet as a figure of the "dark fluidity at the roots of our nature" which "rebels against the security of a *terra firma*",[51] a mark of our own inhuman essence. In this respect, they are not far from the spirit of the *fin-de-siècle* European gentlemen who behold the sun and dream of it turning off.[52]

System cosmology is therefore torn between the Promethean delusions of geoengineers and the intellectual rebellion of nihilists and ecologists. I won't dwell here on this dialectic, which will be the subject of the next chapters of this book. For the sake of my argument, I shall limit myself to point out that the contradiction of System cosmology is a contrast between a hyper-humanist engineerism and an anti- or post-humanist geneticism, that is between an ideological emphasis on fungibility and an ideological emphasis on creativity in relation to information. This could be seen as a mere offshoot of the modern opposition between materialism and spiritualism, if we were not within a paradigm inherently based on the overcoming of this contrast. The passion of objectifying the of Earth is the will to

---

50 Reza Negarestani, 'Globe of Revolution: An Afterthought on Geophilosophical Realism', Journal for Politics, Gender and Culture, 8/2 (2011), 25–54 (p. 25).

51 Nick Land, *The Thirst for Annihilation: Georges Bataille and Virulent Nihilism* (London: Routledge, 1992), p. 107. By the cited authors, see Ray Brassier, *Nihil Unbound: Enlightenment and Extinction* (Palgrave Macmillan, 2007); Eugene Thacker, *In the Dust of this Planet: Horror of Philosophy* (Winchester and Washington: Zer0 Books, 2011); Ben Woodard, *On an Ungrounded Earth: Toward a New Geophilosophy* (New York: Punctum, 2013).

52 Manlio Sgalambro, *La morte del sole* (Milano: Adelphi, 2017), p. 92. In a nihilist cosmology, matter is the death of the spirit (p. 161).

harness an environment,not the instrumentalization of an object; the passion of spiritualizing the Earth is not the constitution of a subject, but the will to give rise to a new form of consciousness.

In conclusion, it is worth spending a few words on what is commonly considered the reference point for contemporary cosmological geneticism: Deleuze and Guattari's geophilosophy. Especially in *A Thousand Plateaus* (published one year after Lovelock's *Gaia*), Deleuze and Guattari claimed that the Earth is more similar to its outside, the Einsteinian universe, than the systemic paradigm allows us to think. Our planet is nothing like a sublimated oasis of life that has grown a spirit, but it too is made of "matter-movement", "matter-energy" or "matter-flow"[53] in a perpetual state of creative divergence. It is simply that this fluid substrate finds a contingent and relative concretion into a texture of terrestrial worldlines apt to further receive and elaborate the universal form-forces. Against the phenomenological Earth as the site of the interiorization of sense, Deleuze and Guattari's Earth (or "Planomene") is an allopoietic system, that is a system whose fluctuating behavior is not aimed at interiorization and stabilization, but at a creative and ungovernable exteriorization. It is not hard to see that this exteriorization ultimately takes as its subject the abstraction of thought.[54] Geophilosophy is basically a neo-Cosmist proposal: a mystical materialism that seeks emancipation in escaping earthly mediations.[55] Deleuze and Guattari are traditionally closer to the accelerationists than to the Gaians.

---

53 Gilles Deleuze and Félix Guattari, *A Thousand Plateaus: Capitalism and Schizophrenia II*, trans. by B. Massumi (Minneapolis and London: University of Minnesota Press, 1987), p. 407.

54 See Peter Hallward, *Out of This World: Deleuze and the Philosophy of Creation* (London: Verso, 2006), especially chapter 7; and Federico Luisetti, *Una vita. Pensiero selvaggio e filosofia dell'intensità* (Milan and Udine: Mimesis, 2011). Luisetti shows the link between Deleuze's ontology of intensive quantities, his philosophy of expression and Kantian transcendentalism. Deleuze's operation is described as the "exhaustion of topology, the abandonment of the natural boundaries of movement and the triumph of illocalization" (p. 21).

55 "Rather than a philosopher of nature, history or the world, rather than any sort of 'fleshy materialist', Deleuze is most appropriately a spiritual, redemptive or subtractive thinker; a thinker preoccupied with the

In taking System cosmology to the extreme, the Planomene really is the last figure of modern cosmology. The allopoietic Earth is the most quintessential transcendental: a purely impersonal field of sensitivity and expressivity towards the Outside—a planet-size *sensorium* inspired to Sartre's "impersonal transcendental field" and Bergson's consciousness as "screen" and complemented with a theory of speculation as creation. The system is precisely this transcendental I definitively purified of the flesh of the empirical subject, freed of all organizations and articulations, of all 'morals', whose creative movement is therefore no less than pure abstraction.

But Deleuze and Guattari also trace a path forward that their nihilist epigones do not undertake. The elimination of all forms of organization enabled by the Planomene is also the catalyzation of new *collective* aggregations along the universal lines of force. In a passage that combines philosophy and poetry, Deleuze and Guattari list some of them:

> (1) pilgrimages to the source, as among salmon; (2) supernumerary assemblies, such as those of locusts or chaffinches, etc. (tens of millions of chaffinches near Thoune in 1950–1951); (3) magnetic or solar-guided migrations; (4) long marches, such as those of the lobsters.[56]

The surreal non-exhaustiveness of this catalog hints towards a political vision that is much wider than that of the Gaians, which may seem to resemble. In chapter 4, I will illustrate the political stakes of Deleuze and Guattari's Planomene in further detail.

---

mechanics of *dis*-embodiment and *de*-materialization. Deleuze's philosophy is oriented by lines of flight that lead out of the world; though not other-worldly, it is *extra*-worldly" (p. 3). Hallward's use of the terms "world" and "materialism" are distinctively modern. In light of the categories outlined in this chapter, Deleuze's position looks already more complicated than this. This passage, however, accurately points out that Deleuze blurs materialism and spiritualism in a specific way fundamentally close to Cosmism. After acknowledging the cosmological nature of Deleuze's philosophy, as Hallward does, one needs to identify which cosmology it endorses and elaborates.

56 Deleuze and Guattari, p. 326.

1.4. *Planet (~)*

The long-term cosmological tendencies I have identified, of course, are not to be intended as homogeneous: for example, they undergo long stages of overlapping and are deeply stratified in the consciousness of people and societies. Furthermore, there are many experiences of thought that deviate from the course of my analysis: counter-tendential cosmologies like Whitehead's anti-Cartesian conceptualization of the World in the frame of a processual Platonic ontology, or Schelling's characterization of the Earth in his early philosophy of nature. Even so, I would maintain that *the cosmic reason of Western modernity is grounded on the problem of co-extensivity between the I and the cosmos*, corresponding to the adventure of the European bourgeoisie in the unfolding of the capitalist world-system. In other words, the problem of totality is tied to the elevation of a social and material viewpoint into a universal form of experience.

The Copernican reconfiguration of the universe was initiated by lower-class craftsmen and sailors, but reflected the experience of another historical subjectivity: the owner of technical means, interested in having an exact and practical knowledge of the material background and the object of their traffics. The exploration of the globe and accumulation of capitals during the 16th and 17th centuries required an artificial access to reality which reproduced the controlled observation of the experimental method. The use of technological devices became fundamental not only to discover reality, but to construct a *locus mundus*: a discernible, neutral space accessible to the mind. The knowledge of the world gradually separated itself from praxis and became an abstract power of reason.

Mechanicism as the ideology of manufacture—as it is often and rightly depicted—was therefore an outgrowth of the mathematized world of merchant capitalism. As manufacture gave way to industry, the exploration of the world gave way to the appropriation and domestication of the Earth. At the heart of modernity, "man enters into possession of the earth's bowels, traverses and labors, trespasses onto what had previously been unreachable".[57] The regime of experience corresponding to this new subject is a universalizing projection of the subjective structure itself. If World cosmology was a cosmology of separation, Earth cosmology was a cosmology of reconciliation—between reason and morality, knowledge and power—in light of the triumphant subject of modernity.

Earth cosmology as well, however, was marked by an instability that soon materialized into a crisis. System cosmology was the new compromise after that crisis, and as such was aimed at producing a wider reconciliation between knowledge and power based on an enlarged notion of experience: not the experience of the subject (bearer of the old idea of 'humanity'), but the functioning of the system coupled with its environment. Neither mechanical nor organic, the system was this new integrator of contradictions—as Norbert Wiener, the father of the informationalist paradigm, clearly spelled out.

The system itself was no less unstable than the previous frameworks. On one hand, the systemic reason draws an analogical continuity that goes from the individual to the planet. The human individual becomes a system among systems. On the other hand, however, the individual subject is no longer at the center or at the pinnacle of an oriented development that goes from the cosmos to humanity. All that remains is the 'self' of the system's self-regulation. Every system can be a subject as much as it can be an object; every whole can be subjectified and objectified from the viewpoint of another. It is Heidegger's shock in front of the first photographs of the Earth from space, the terrible realization of being able to see God there in front of you. A cosmos was built that lay outside of the control of its inhabitants. The knowledge of the world became paranoia.

The crisis of the modern subject as a cosmological apparatus of commensurability and continuity with totality left a vacuum that the system cannot fill due to its intrinsic contradictions. Is the 'self' of the Earth an *autós* or an *állos*? The conceptual structure of modern subjectivity served precisely to solve this dichotomy. In

57 Gilbert Simondon, *On the Mode of Existence of Technical Objects*, trans. by C. Malaspina and J. Rogove (Minneapolis: Univocal, 2017), pp. 141–142.

the former case, as an *autós*, the Earth is the outcome of a process of identification, and can therefore be objectified and manipulated. This conclusion bears unacceptable consequences on both the theoretical and the practical level. In the latter case, instead, the Earth is conceptualized as a process of alteration. I contend that either as an alteration towards the cosmic environment or as an internal pluralization, the anti-systemic Earth remains a dangerous abstraction that brings about no convincing answers to the question of experience.

In what I have described as the contradiction of systemic cosmology lie the seeds of its dephasing. As I will claim in the next chapters, the internal contrast between accelerationists and Gaians has already given birth to the cosmological figure that may lead to the coming cosmology: the planet. Right now, we find ourselves in the middle of this unstable subphase of the third cosmology. Every modern cosmology has undergone a similar phase: the World gave way to the world as a transitional solution before the paradigm shift to the Earth; the Earth gave way to the ark before passing into the System. Now System cosmology finds itself in a similar state of imbalance. We live in an epoch of controversy, a morbid interregnum, and the planet is the figure of our cosmological uncertainty.

Cosmology	Ideology	Contradictions	Dephasing
World (17th–18th century)	Mechanicism	Rationalism / Empiricism	world
Earth (18th–20th century)	Vitalism	Spiritualism / Materialism	Ark
System (20th–21st century)	Informationalism	Engineerism / Geneticism	Planet

2.
PLANET CULT

2.1. *The need for a new universalism*

A discourse on the 'planet' began to emerge together with what I defined as the third cosmology of modernity over a timespan that goes from 1945 to the first half of the 1970s. At first, it was in fact barely distinguishable from the incipient cosmological paradigm. Right after World War 2, the planetary idea was simply in the air. Intellectuals, policy-makers and investors from different contexts began to underscore the necessity of inventing political and legal formations that could enact regulations on a global scale, beyond the party interests of the new global contenders, the US and the URSS. A certain emphasis on technology and anthropology was usually key to evoke the overcoming of the opposing ideologies: the world, it was argued, must be united due to the planetary scale of human action on reality and the risks coming from its unregulated use.

As an ideology in formation (that is a super-ordinate and often contradictory aggregate of beliefs and positions), the discourse on world unification appeared here and there and migrated, retaining only some basic coordinates. For this reason, it can only be approached empirically, starting from its different instantiations. Here is one:

> The great problem of the world as it already exists calls for a solution: How shall this archaic mammal, with its archaic needs that have been the driving force of its ascent, continue to push its rock up the hillside if one day it is left with only the image of its reality? At no time in its development has this species yet had to break away from itself since the days of the Australanthrope. *Homo sapiens* lived his interminable adventure concretely; today the human is on the point of exhausting the

> resources of the planet, and already the myth of human transplantation into space has sprung up. But there can be no going back over the ground already covered. We can dream that when arriving on a distant star, the human will encounter Pithecanthropus and the southern elephant but will not revert to flint knapping.
>
> [...] We can imagine the human of the near future as being determined by a new awareness and the will to remain *sapiens*. In such an event the problem of the individual's relationship with society will have to be completely rethought: We must face up squarely to the question of our numerical density and our relations with the animal and plant worlds; we must stop miming the behavior of a microbic culture and come to grips with the management of our planet in terms other than those of a game of chance.[1]

Thus ends André Leroi-Gourhan's *Gesture and Speech*, published in 1964, with an anthropology of the planetary future exemplifying the epoch's state of consciousness. The French paleo-anthropologist offered few answers, but raised the problem with rare clarity: humanity is witnessing a fundamental and accelerated cosmological change, a break comparable only to the invention of agriculture and urbanism in the Neolithic, which will steer the course of hominization quickly and deeply. This shift in technology and material relations, however, still lacks a culture that can frame and absorb it at the level of values. "Today, despite efforts being made in the social sphere and despite decolonization, the planetary human group still has the same form as the small Mesopotamian societies of four thousand years ago".[2] Although everything, from transport infrastructures to the way people dress, hints at "the transition to a humanity different from the one still remembered by living generations",[3] our cosmology is still set on the model of the Neolithic settlement.

Leroi-Gourhan's were the words of a Catholic European intellectual who found himself in between two emerging planetary societies, the anthill-society of URSS and the individualistic society of the US. Leaving "somehow undecided the ultimately materialist

1 André Leroi-Gourhan, *Gesture and Speech*, trans. by A.B. Berger (Cambridge, MA, and London: MIT Press, 1994), pp. 407–408.
2 Leroi-Gourhan, p. 184.
3 Leroi-Gourhan, p. 352.

or idealist tenor of his claims",[4] he posed the problem in terms of a new anthropological ecumenism dictated by the sheer scale of humanity's common problems. Post-war progressive Catholic culture was indeed a fertile environment for this kind of message during a period of heightened Christian ecumenism.[5] Many believed that "a world straining for unification but isolating itself into separate sectors and local entities, sometimes out of national solidarity, sometimes on a continental scale"[6] demanded new valued which the Church, with its centuries-old universalist aspirations and metapolitical positioning, could provide. New transnational formations like the UN and UNESCO had already begun to spread the values of a depoliticized globalism; French anthropology held a certain role in that process too.[7]

Jacques Maritain—one of the most prominent intellectuals in French Catholicism, head of the French delegation at UNESCO and member of the committee for the Universal Declaration of Human Rights—was exemplarily convinced that a peaceful and just world had to be built upon universal human rights and the creation of a global community endowed with decisional powers and founded on the abandonment of the concept of national sovereignty. Following his pre-war commitment to a "new Christendom" inspired by Thomistic political philosophy, he spread his convictions in a wide number of official occasions and influential circles. His ideas also inspired the work of the Committee to Frame a World Constitution and its manifesto, the *Preliminary Draft of a World Constitution* (1948). The leaders of the committee, Maritain's friend Mortimer Adler (an eclectic and peculiar figure of philosopher and educator) and Robert Maynard Hutchins (Chancellor of the University of Chicago), were perfect examples of how the Catholic message

4 Nathan Schlanger, 'Between Endorsement and Disavowal: André Leroi-Gourhan's Russian Interactions', Technology and Language, 5/2 (2024), 136–152 (p. 139).

5 See Massimo Faggioli and Bryan Froehle, *Global Catholicism: Between Disruption and Encounter* (Leiden and Boston: Brill, 2025).

6 Yves Congar, *Ecumenism and the Future of the Church* (Chicago: The Priority Press, 1967), p. 174

7 See Bernard D. Gheoghegan, *Code: From Information Theory to French Theory* (New York: Duke University Press, 2023), chapter four.

found fertile ground in a stratum of religious intellectuals of the new America. In 1949, Hutchins delivered a lecture on St. Thomas and the World State, arguing that "according to the mind of St. Thomas, only the world state can now be the perfect community".[8] During his years in Princeton and Toronto, Maritain will develop a strong faith in the future role of the US in building the global Christendom.

Another French Catholic intellectual of the time, Pierre Teilhard de Chardin, is today considered as the forefather of planetary philosophy. Although more visionary in their tone and far more heterodox than Maritain's, Teilhard's convictions were the same: "The age of nations has passed. Now, unless we wish to perish, we must shake off our old prejudices and build the earth".[9] Like Maritain (and even to a greater extent), Teilhard exerted a key influence in the constitution of the United Nations.[10] Unlike Maritain, however, his arguments in favor of a world state were sustained by a complete philosophy of planetary consciousness.

> It is an amazing thing that in less than a million years the human species has succeeded in covering the earth, and not only spatially. On this surface that is now completely encircled, mankind has completed the construction of a close network of planetary links, so successfully that a special envelope now stretches over the old biosphere. Every day this new integration grows in strength. It may be clearly recognized and distinguished in every quarter. It is provided with its own system of internal connections and communication, and for this I have for a long time proposed the name noosphere.[11]

The concept of "noosphere", a term borrowed by Teilhard from Vladimir Vernadskij as early as 1922, described the planetary

8 Robert Maynard Hutchins, *St. Thomas and the World State* (Milwaukee, WI: Marquette University Press, 1949), p. 44.

9 Pierre Teilhard de Chardin, *Human Energy*, trans. by J.M. Cohen (New York: Harcourt, 1969), p. 37.

10 See Sarah Shortall, '"Building the Earth": Pierre Teilhard de Chardin, Science, and the Spirituality of the United Nations', Journal of the History of Ideas, 85/4 (2024), 827–855.

11 Pierre Teilhard de Chardin, *Activation of Energy*, trans. by R. Hague (New York: Harcourt, 1970), p. 285.

interaction of human minds. "The more scientifically I regard the world", declared Teilhard, "the less can I see any possible biological future for it except the active consciousness of its unity".[12] A scientific diagnosis of the state of things led him to the inevitable conclusion about the necessity of a spiritual unification on a planetary scale entailing a leap up, a jump of the species—or a "conversion".[13]

At the roots of all these three thinkers (Leroi-Gourhan, Maritain, Teilhard) was the philosophy of Henri Bergson, who in the *Two Sources of Morality and Religion* first prophesized a planetary mysticism built upon a Positivist idea of ecumenism as universal love for humankind. Planetary universalism can indeed be associated with a Bergsonian *morale ouverte*.[14] However, none of these ideas would have led to a full-fledged cosmology if it had not been for the further consequences of the change in technological paradigm sensed by these authors. Between the 1950s and '60s the emergence of the Net, the birth of the first global media and the infrastructural

---

12 Teilhard de Chardin, *Human Energy*, p. 37.

13 "*A new plane* (individual, social and religious) *must eliminate another.* This entails inner tortures and persecutions. *The earth will only become conscious of itself through the crisis of conversion*" (Teilhard de Chardin, *Human Energy*, p. 38).

14 The line that goes from Bergson to the first planetary ideologues in France still deserves detailed investigation. Here I just quote the most telling passages from the *Two Sources*, which already say much about the deep roots of this conceptual constellation. "Let a mystic genius but appear, he will draw after him a humanity already vastly grown in body, and whose soul he has transfigured. He will yearn to make of it a new species, or rather deliver it from the necessity of being a species [...]. The great breath of life which swept our planet had carried organization as far along as nature, alike docile and recalcitrant, permitted". "Mankind lies groaning, half crushed beneath the weight of its own progress. Men do not sufficiently realize that their future is in their own hands. Theirs is the task of determining first of all whether they want to go on living or not. Theirs the responsibility, then, for deciding if they want merely to live, or intend to make just the extra effort required for fulfilling, even on their refractory planet, the essential function of the universe, which is a machine for the making of gods" (Henri Bergson, *The Two Sources of Morality and Religion*, trans. by R.A. Audra and C. Brereton with the assistance of W.H. Carter [Westport, Conn.: Greenwood Press, 1974], pp. 300, 306).

revolution in logistics concretized some of the most futuristic visions uttered in the decades before.

It is no a chance, then, that the first true prophets of the planetary appeared in the US, home of cybernetics and the post-war technological revolution at large. In 1964, while the gigantic globe of the Unisphere (the biggest iron monument in the world) was becoming the symbol of the New York World's Fair, Marshall McLuhan, a converted Catholic, illustrated his ideas at the American Cybernetic Society and argued that, as humankind was called to "cope with abundances and riches in both the economic and psychic order"[15] like never before, the "electric age" was repatterning human nature. The advent of electricity and then the birth of cybernetics "made possible the extension of the human nervous system as a new social environment",[16] an integrated but decentralized milieu made of pure information. They were living in an epoch comparable only to the anthropological revolution of the Neolithic, and in the near future, all human cultures would be reunited "as responsive cybernetic systems"[17] into a hyper-system resembling a global collective consciousness.

McLuhan was soon to be acclaimed as an oracle of modern times. Whereas a humanist like Leroi-Gourhan acknowledged both the evolutionary preeminence of technology and the capacity of culture to integrate and harmonize it, McLuhan seemed not to believe that human culture could match technology in speed. In his prophecies, culture was reduced to a coping reflex in front of the revolutions of technology and therefore confined into a position of defense of psychological and social integrity. "We are suddenly forced to live in such complex and compressed and high-speed systems", McLuhan argued, "that we inevitably switch into mythic patterns".[18] Present inadequacy meant regression. Even though in some passages

15 Marshall McLuhan, 'Cybernetics and Human Culture' (1964), in *Understanding Me: Lectures and Interviews*, ed. by S. McLuhan and D. Staines (Toronto, 2003), pp. 44–55 (p. 55).

16 McLuhan, p. 49.

17 McLuhan, p. 48.

18 Marshall McLuhan, 'The Future of Man in the Electric Age,' in *Understanding Me*, 56–75 (p. 63).

McLuhan seemed to believe that great cultural movements like "the ecumenical movement or the liturgical movement of our time" could positively mobilize the coeval societies, the dominant tone of his work was one of threat and resignation towards a collective involution.

McLuhan depicted a situation of chaos and conflict (a return to the state of nature)[19] and a drama of human identity only to underscore with greater intensity the necessity of a new—and in fact very old—system of values. In later works, this perspective emerge more clearly: the tragic contrast between technology and culture urges the "man" to become shepherd of himself, or rather—in more evangelical terms—"his brother's keeper, in spite of himself".

> The role of the shepherd, a continuing archetype in biblical literature, invariably entails a spiritual quest. The wolf prowling ravenously about the flock is God's agent for self-examination. In the near future, our spiritual quest may lead us not, as some have thought, into outer space but into the depths of the sea for additional sources of food and different techniques of survival. But such an exploration of inner space could, at first, take the form of a radical overhaul of educational methods.[20]

Humanity must reprogram its own system of values and assume the role of steward of the humanized planet, that is to say the Westernized global village. This task—he maintained—fundamentally passes through education.

McLuhan's pessimist moralism stemmed from a harsh vision of modernity as the triumph of Protestantism over Catholicism and underlaid a conservative ideal centered on old-times values, "all

---

19 This is a common misunderstanding of McLuhan's view: "retribalization" is not at all a peaceful condition. In a 1977 interview, he precised: "Tribal people, one of their main kinds of sport is butchering each other. […] When people get close together, they get more and more savage, impatient with each other. […] Village people aren't that much in love with each other. The global village is a place of a very arduous interfaces and very abrasive situations" (Marshall McLuhan, 'Violence as a Quest for Identity', in *Understanding me*, 264–276 [p. 265]).

20 Marshall McLuhan and Bruce R. Powers, *The Global Village: Transformations in World Life and Media in the 21st Century* (New York: Oxford University Press, 1989), p. 100.

those merely human faculties which produce games and philosophy, and poetry and music and mirth and fellowship with a very fleshy basis".[21] What at first sight might look like a pessimist diagnosis of the present was in fact a nuanced call for an old-fashioned communitarian moral; what might seem a difficult, pondered optimism was in fact a defensive critique of secularization.

The French Catholic milieu and McLuhan's theories are two relevant instantiations of a much wider rhetoric. Since 1945, the planetary discourse sustained a great number of initiatives promoted by a heterogeneous assortment of intellectuals, like the United World Federalists, Inc., endorsed by Albert Einstein; the People's World Convention (1950–51), sponsored by Einstein, Thomas Mann and Robert Hutchins among others, and the People's World Constituent Assemblies that followed it throughout the next decades; the World Constitution Coordinating Committee (1957); the World Committee for a World Constitution (1958); later, the Consultative Council to the World Constitutional Convention (1967) and the World Constitutional Convention (1968). The idea of the political or at least institutional and legal unification of the world came out of the great constituent momentum that followed the end of World War 2, which will run out during the 1950s as the Cold War tightened its grip on international politics.

Let us take one last example of how transversal the planetary rhetoric initially was. On the other side of the Iron Curtain, the futurological writings of Polish sci-fi novelist Stanisław Lem (himself an attentive reader of Thomas Aquinas and Bergson) reverberated the combination of spiritualism and technological determinism that characterized the post-war imagery. As a writer, Lem was in the position of manipulating this new imagery with both faith and skepticism. He wrote in 1963: "We seem to be at the end of an era.

21 Marshall McLuhan, *Letters* (New York: Oxford University Press, 1987), p. 72. He continues: "Everything that is especially hateful and devilish and inhuman about the conditions and strains of modern industrial society is not only Protestant in origin, but it is their boast to have originated it" (p. 73). On McLuhan conservatism see Grant Havers, 'The right-wing postmodernism of Marshall McLuhan', Media, Culture & Society, 25 (2003), 511–525.

[…] Human civilization is like a ship that has been built without any design plans. […] Civilization lacks knowledge that would allow it to choose a path knowingly from the many possible ones, instead of drifting in random tides of discoveries". Therefore, "we have to learn how to regulate scientific progress too; otherwise, the random character of any future developments will only increase". "If I say that one era is coming to an end, I am not even thinking about the demise of the old systems. Satisfying the basic needs of humanity is a necessary task, a preparation for a final exam; it is the beginning of a mature age rather than its end".[22] The problem that prevented humanity from facing its existential needs was disorder, randomness, conflict. The solution lay in the cultural elaboration of knowledge, which—it was implied—would automatically lead to social reform.

The fact that Lem made no reference to past or present history[23] is the most evident sign of his participation in the rhetoric of planetary universalism. His approach can be traced back to the growing discontent towards *socrealizm* and Stalinism, heightened after Polish October in 1956,[24] but can also be interpreted in relation to the state of cybernetic knowledge in the URSS. Between the 1950s and the '60s, the Soviet Union had absorbed and magnified cybernetics as a post-Stalinist techno-utopia. The level of techno-scientific knowledge and the degree of penetration into popular culture had

---

22 Stanisław Lem, *Summa Technologiae*, trans. by J. Zylinska (Minneapolis and London: University of Minnesota Press, 2013), pp. 235–236.

23 As it has rightly been noted, "on his way to stating that the ship does not know where to go, Lem blithely eliminates 4,000 years of war, colonialism, and exploitation—no wonder the course is not set straight. And yet, as so much of the science fiction that Lem loved to deride suggests, perhaps this 'something' is precisely why the ship cannot sail on" (Paweł Frelik, 'Stanisław Lem's *Summa Technologiae* as Impossible Utopia', Science Fiction Studies, 40/3 [2013], 439–450 [p. 441]). Yet Lem was far from disinterested or ignorant in history and politics. The commentator's hypothesis is that Lem leant towards a sort of scientist idealism, whose degree of abstraction allowed him to start from the same premises as more spiritualist thinkers than him.

24 See Carl Tighe, 'Stanisław Lem: Socio-Political Sci-Fi', The Modern Language Review, 94/3 (1999), 758–774.

become comparable and at times higher than in the US.[25] But the dream of a cybernetic reformation of communism did not take root in the Soviet establishment, as became evident with the succession from Khrushchev to Brezhnev at the top of the party in 1964. At the beginning of the 1960s, the imagination of communist Russia was going around in circles.

Lem's vision was similarly contested between utopianism and diffidence. Differently from the European spiritualists like Teilhard (which he openly criticized), the atheist Lem accepted the idea of planetary civilization in a more exploratory and less eschatological manner. In all its skepticism, however, his approach ultimately underlay the same mix of rationalism and mysticism around technology which would have soon become the mark of the epoch. "I do not trust any promises", he wrote, "nor do I believe in any assurances built on so-called humanism. The only way to deal with technology is with another technology. Man knows more about his dangerous tendencies than he did a hundred years ago, and in the next hundred years, his knowledge will be even more advanced. Then he will make use of it".[26]

2.2. *Los Angeles is the future for us all*

In the US, the initiatives for world unification took on a more precise political and social identity after the consolidation of the Cold War scenario. "World order" became the constructive side of the politics of resistance to a possible Soviet aggression,[27] and post-

25 See Richard Barbrook, *Imaginary Futures: From Thinking Machines to the Global Village* (London and Ann Arbor, MI: Pluto Press, 2007), chapters 10 and 11.

26 Lem, p. 6.

27 See Stanley Hoffman, *Primacy or World Order: American Foreign Policy since the Cold War* (New York: McGraw-Hill 1978). The world order politics was publicly recognized in the declarations of presidents such as Henri Kissinger and Jimmy Carter. Kissinger will later write: "Imbued with the conviction that its course would shape the destiny of mankind, America has, over its history, played a paradoxical role in world order: it expanded across a continent in the name of Manifest Destiny while abjuring any imperial designs; exerted a decisive influence on momentous

war "idealistic activism"[28] took more clearly the guise of a deliberate political universalism. In 1961, Hermann Kahn, one of the fathers of the Cold War theory and modern futurology, co-founded the Hudson Institute, a think tank devoted to the strategic extension of the US hegemony to the globe. Three years later, the Institute for World Order was founded with the task of providing "an analytic, holistic, and ethical framework for the study and promotion of a just world order".[29] During the symposium held in Bellagio in 1965 and titled *Conditions of World Order*, French philosopher Raymond Aron (who served as an early advisor for the Hudson Institute) provided a definition of world order as a minimum requirement for coexistence: the question was "under what conditions would men (divided in so many ways) be able not merely to avoid destruction, but to live together relatively well in one planet".[30]

Gatherings and institutions such as these were the public face of a network that connected the US government with intellectuals and influential private actors (such as the Rockefellers) around the

events while disclaiming any motivation of national interest; and became a superpower while disavowing any intention to conduct power politics. America's foreign policy has reflected the conviction that its domestic principles were self-evidently universal and their application at all times salutary; that the real challenge of American engagement abroad was not foreign policy in the traditional sense but a project of spreading values that it believed all other peoples aspired to replicate" (Henri Kissinger, *World Order: Reflections on the Character of Nations and the Course of History* [New York: Penguin Books, 2014]).

28 Hoffman, p. 199.

29 Saul H. Mendlovitz, 'The Program of the Institute of World Order', Journal of International Affairs, 31/2 (1977), 259–265 (p. 259). The first phase of the Institute's activities (1961–66) culminated in the publication of the collection *The Strategy of World Order: Toward a Theory of War Prevention*, edited by Richard Falk and Saul Mendlovitz and financed by the World Law Fund, and the foundation of the 'World Order Studies'. Since then, the Institute has devoted its work to a futurology of 'preferred worlds'. See Richard A. Falk, 'Contending Approaches to World Order', Journal of International Affairs, 31/2 (1977), 171–198.

30 Stanley Hoffman, 'Report of the Conference on Conditions of World-Order – June 12–19 1965, Villa Serbelloni, Bellagio, Italy", Dedalus, Spring 1966, 455–478 (p. 456).

common goal of an American planet. During the 1950s, this task was organically taken on by the so-called "Cold War Left", a movement of progressive intellectuals whose main endeavor became that of crafting a "post-Fordist social utopia for the American empire".[31] As argued by Richard Barbrook, they found the principles of this new vision in the prophet of the Net, Marshall McLuhan. Although tinged with Christian mysticism and conservatism, McLuhan's oracles were perfectly in line with the hopes and dreams of the new America: "Technology was the driving force of human history. The convergence of computing, media and telecommunications into the Net was creating a new social system. The production of goods was being supplanted by the provision of services. Representative democracy would soon be supplemented by on-line voting. The nation state was being integrated into the process of world unification".[32] The first manifesto of this operation, *Between Two Ages: America's Role in the Technetronic Era* (1970) by member of the Bell Commission Zbigniew Brzezinski, openly indicated in planetary consciousness the ultimate horizon of this prophecy.[33]

Authors like Brzezinski and Daniel Bell turned the predictions on the planetary age into concrete description of the upcoming society. Planetary ideology, however, still required its grand narrative. The history of a now forgotten gathering of intellectuals called the Lindisfarne Association is particularly instructive in this

31 Barbrook, p. 147. On the Cold War Left, see chapter 7.

32 Barbrook, p. 175.

33 Brzezinski explicitly entrusted this vision to the hands of new transnational elites reunited by the cybernetic grid (see Zbigniew Brzezinski, *Between Two Ages: America's Role in the Technetronic Era* [New York: The Viking Press, 1970], pp. 28–30). The historical subject in charge of embracing the effects of the new technologies was clearly indicated in the most dynamic stratum of the productive class, that which James Burnham had described as the new managerial élite (see James Burnham, *The Managerial Revolution: What is Happening in the World* [New York: John Day Company, 1941]. As explained by Barbrook, "James Burnham was the pioneer of the move by left-wing intellectuals from the margins into the mainstream of US society", that is one of the intellectual fathers of the Cold War Left (Barbrook, p. 92). For these former Marxists hired by American liberalism, the economic factor—largely ignored by the first planetary ideologues—was decisive.

sense. The Lindisfarne Association was founded in 1972 by William Irwin Thompson, then a professor at York University in Toronto and an author of countercultural best-sellers. The Association's declared mission was studying and realizing a planetary culture. The material opportunity of creating the Association was presented to Thompson by a passionate advertising executive, Gene Fairly, who raised enough money to knock on the door of the Rockefeller Brothers fund. Thompson recalled the atmosphere of electric excitement around the theme of planetarity in the New York high society of the epoch. The Lindisfarne Association was

> a seventies kind of cultural movement in that it sought to avoid the sixties-style mass movements of drugs and revolutionary violence to seek out a more spiritual and intellectual third way to effect the transformation of culture. As a student of cultural history, I had a sense that our Western civilization was at the edge of its kairos and that a newly emerging planetary culture called out for a new formation with which to realize itself, so I sought out others in what became the Lindisfarne Fellowship that seemed to embody this third way.[34]

Thompson capitalized on this epoch-defining sentiment to create a syncretic ideology based on post-religious spirituality, ecological communitarianism and cybernetics. The mission of Lindisfarne was to produce and diffuse the new planetary knowledge through "a new kind of educational community in which the individual was empowered through meditation to connect the *unique* to the *universal* without the mediation of clerical ritual and collectivizing worship".[35] The Association soon became a center of attraction for like-minded intellectuals, with annual conferences held from 1974 to 1977, and an attempt to create a 'glocal' way of life within informatically

---

34 William Irwin Thompson, *Thinking Together at the Edge of History: A Memoir of the Lindisfarne Association, 1972-2012* (Traverse City, MI: Lorian Press, 2016), p. 1.

35 Thompson, p. 9.

connected communities of subsistence according to the principle of New Communalism.[36]

Before founding the Association, Thompson had been a receptive witness of the spread of planetary cosmology in Californian counterculture. Taking very seriously the mystical overtones of McLuhan's doctrine (whom he met at MIT in the 1960s), he saw in Los Angeles—with its "violent juxtaposition of aerospace technology and neo-tribal politics"—"the future for us all".[37] After the end of modernity, he argued, the dramatic contrast between cultural particularity and technological universality was preparing a new synthesis that would occur in the 21st century, when a "Scientific-Planetary Civilization" will arise after a phase of disaster and chaos. The Californian "children of McLuhan",[38] specimens of a new humanity, were the chosen people that would later emerge triumphant from the apocalypse. Indeed, they would soon be the front line of the new neoliberal élite.

The Lindisfarne Association participated in this process at a purely ideological level. The first annual conferences showed a common commitment to intellectually naturalizing decentralization and global unity; prophecies were pronounced about the disappearing of multinational corporations as "the most serious crisis of all",[39] plans of world order were made based on "the notion of benefit for all humanity"[40] to be spread through education. Malthusian

---

36 On New Communalism, see Fred Turner, *From Counterculture to Cyberculture: Stewart Brand, The Whole Earth Network, and the Rise of Digital Utopianism* (Chicago and London: The University of Chicago Press, 2006), chapter 1.

37 William Irwin Thompson, *At the Edge of History: Speculations on the Transformation of Culture* (New York, Evanston, San Francisco and London: Harper & Row, 1972), p. 19.

38 Thompson, *At the Edge of History*, p. 46.

39 Richard Barnet, 'Global Political Alternatives', in Michael Katz, William P. Marsh, Gail Gordon Thompson (eds.), *Earth's Answer. Explorations of the Planetary Culture at the Lindisfarne Conferences* (New York: The Lindisfarne Association / Harper & Row, 1977), 114–121 (p. 121)

40 Saul Mendlovitz, 'The Consciousness of Evolution', in Katz, Marsh, Thompson, 122–133 (p. 131). The abstraction of such notion of "humanity" is explicitly thematized and vindicated: "If you invited the blacks and the

demiurges pondered over the fact that, if they had brought peace and civilization to the whole world, they would also raise a problem of overpopulation.[41] Gregory Bateson, called to apply his theories to the social field, argued for the necessity of calculating the changes in human evolution on a long term and at the level of the "total adaptive system".[42]

Lindisfarne counted among his affiliated fellows prominent figures of the American system sciences such as Bateson (regarded as a spiritual mentor of the Association), Francisco Varela, David Abram, Lynn Margulis, James Lovelock and Stuart Kauffman.[43] By the time of the first annual conference, system science had become the field of synthesis dreamt by cybernetics. Wrapping up the 1975

Third Worlders", says the speaker at the beginning of his talk, "they probably wouldn't come. They don't want to be here; it's not their thing. [...] Perhaps we should make a virtue of it by trying to understand what, if any, relationship we might usefully have with them" (p. 124).

41 "We are speaking [...] about a dangerous game. [...] The Rockefeller Foundation, for example, with its humanitarian impulse and great beneficence, caused malaria to be eliminated in the Mauritius Islands, reducing infant and child mortality from this disease. The consequences in terms of the population explosion and the misery that followed are well known. Similarly, we are altering the human gene pool through acting out our humanitarian impulses. At some point, we will have to re-examine our values and priorities and begin to look at ourselves from an evolutionary viewpoint. We are at a moment of crisis of danger and opportunity" (Jonas Salk, 'The Consciousness of Evolution', in Katz, Marsh, Thompson, 134–141 [p. 136]).

42 Gregory Bateson, 'The Thing of It Is', in Katz, Marsh, Thompson, 142–154 (p. 154). After all, the acclaimed anthropologist was accustomed to craft epistemological tools for surveilling and managing the social body: first the colonial cultural system in Bali, then the family unit in postwar America (see Gheoghegan, chapter 2), now the planet.

43 "When Lindisfarne's ecological world view was enhanced by the critique of Computationalism in the 'Embodied' cognitive science of Francisco Varela in 1977, and then by the Gaia evolutionary theory of James Lovelock and Lynn Margulis in 1981, it began to be obvious to all of us that a new science was showing its face at Lindisfarne and that just such a science was as critical to the process of planetization as any esoteric philosophy of the past" (Thompson, *Thinking Together at the Edge of History*, p. 34).

conference, with constant references to what just said by Bateson (called "Gregory" with ostentatious familiarity), Thompson warned that if they wanted to steer the course of human evolution in a planetary sense, they would have to express a new universalism for both the masses and the élites (a purpose, we can infer, that had always been behind the religious nature of his vision). Thompson's own solution was a mix of process philosophy, system science and mystical anti-modernism, as embodied in the project of self-sufficient ecological communities carrying on the "resacralization of humanity".[44] This future-oriented, techno-mystical, esoteric, holistic humanism was already, in Thompson's own words, a "posthumanism".[45]

During those years, planetary ideology migrated back and forth the Atlantic. Especially in France, which was tied to the US by common intellectual trends,[46] this rhetoric found fertile ground. An interesting figure of French third-way intellectual[47] and planetary ideologue was Joël de Rosnay. Son of a descendant of colonialists in the Mauritius and a fugitive from the Russian Revolution, later married to the daughter of a former secretary general of the UN, de Rosnay made career in between science and business. One can find him participating in crucial junctures like the Groupe des dix, a think tank formed in 1969 with the aim of applying system sciences

---

44 William Irwin Thompson, 'Notes on an Emerging Planet', in Katz, Marsh, Thompson, 206–229 (p. 225).

45 William Irwin Thompson, *Passages About Earth. An Exploration of the New Planetary Culture* (New York, Evanston, San Francisco and London: Harper & Row, 1973), p. 140.

46 A transatlantic intellectual field formed between the US and Western Europe (and France in particular) around the sociological paradigm of "industrial society", championed by authors like Bell and Aron (see Hammond David Sessions, *Man, Machines, and Modernity: Inventing "Industrial Society" in French Sociology 1930–1981*, PhD Dissertation, Boston College, 2021). Aron was somewhat of a symbol of this osmosis, as he participated deeply in the cultural scene of post-war US. The association between planetary cosmology and the industrial society paradigm—with its pendant, the notion of "post-industrial society"—is a fundamental element of the story I am retracing here which I do not have time to analyze.

47 Barbrook, p. 164.

to society, together with other key intellectuals of the time such as Edgar Morin, Henri Laborit and Henri Atlan.

One of de Rosnay's most known books, *The Macroscope* (1975), was a hymn to the rising tide of planetarization. The rhetoric incipit was always the same: "The explosion in the means of communication, the acceleration of evolution, and the interplay of energy and economics have torn away the veil that hid the planetary totality from our eyes. [...] Through 'world public opinion,' 'collective consciousness'—and even the 'collective unconscious'—we see the outline of an emergent, psychology, of the noosphere gradually taking shape".[48] But post-war utopianism had already given way to a new awareness of the limits of planetary totality: energy crises and material scarcity, natural catastrophes, social turmoil and economic oscillations, violence—in a word, instability, the old enemy of planetary unification—were exerting a both unifying and delimiting effect. In the mid-1970s, the world order project was already a partial balance of "failure".[49]

For de Rosnay, the political responsibility still lay on the shoulder of a 'third force' between "unrestrained capitalism" and "bureaucratic communism", a force which would work towards an open and pluralistic "ecosociety". The problem had become "how to reorient the economy to serve better at the same time human needs, the maintenance and evolution of the social system, and the pursuit of true cooperation with nature". This was the form that global capitalism had to assume to evolve beyond Fordism: a form aimed at "controlled disequilibrium".[50] Precisely in this sense could system sciences provide a descriptive frame for society. By cultivating and spreading this "new culture", argued de Rosnay, it would be soon possible to transform the basic principles of authority, work, politics and rationality itself on a planetary scale. Ten years later, de Rosnay will re-assert and update his convictions: "We are the neurons of the Earth: the cellules of a forming brain of planetary dimensions. Despite our struggles, and our hatreds,

48 Joël de Rosnay, *The Macroscope: A New World Scientific System*, trans. by R. Edwards (New York, Hagerstown, San Francisco and London: Harper & Row, 1979), p. 194.
49 De Rosnay, p. 204.
50 De Rosnay, p. 222.

despite the chasm of our differences we all participate, consciously or unconsciously, in this construction on which our destiny depend".[51]

The ideologues of systems science will keep cultivating the discourse on planetary culture well into the 1980s and the decades to come. In 2008, an old Stuart Kauffman still wrote: "We lack a global ethic. We must construct one to prepare for the global civilization that is upon us, hopefully persistently diverse and forever evolving, forever inventing, and partially guide the form it will take".[52] Planetary civilization has apparently remained impending for seventy years. In 2014, best-selling physician Fritjof Capra and chemist Pier Luigi Luisi published a monumental summa of systems philosophy culminating in a slightly less idealized project to "reshape globalization",[53] that is to reform global capitalism. The dreadful suspicion that these authors, by continuing to speak the language of governance and ethical responsibility, refuse to face is that the planet has always been first and foremost a political project.

2.3. *Whole Earth paranoia, geodesic Futurism and the radical design of life*

During the Cold War, the discourse on planetary culture reflected the changing relation between progressive intellectuals and the international politics of the West guided by the US. But planetary thinking was also the expression of wider cultural movements that were reprogramming the boundaries of experience and catalyzing collective imagination. The idea of the planet, which as I said was 'in the air', emerged directly from the social unconscious.

Often, the most relevant figures of planetary ideology were not even traditional intellectuals. Take Stewart Brand. Graduated in biology at

51 Joël de Rosnay, *Le cerveau planetaire* (Paris: Olivier Orban, 1986), p. 11.

52 Stuart A. Kauffman, *Reinventing the Sacred: A New View of Science, Reason, and Religion. The Science of Complexity and the Emergence of a Natural Divinity* (New York: Basic Books, 2010), p. 273.

53 Fritjof Capra and Pier Luigi Luisi, *The Systems View of Life: A Unifying Vision* (Cambridge, UK: Cambridge University Press, 2014), chapter 18.1.1.

Stanford, then a student in design and photography in San Francisco, in 1966 he led a public campaign named "Why haven't we seen a photograph of the whole Earth yet?". NASA was rumored to possess photos of the whole Earth from space, and Brand—reportedly inspired by an LSD trip[54]—believed that these images could become a symbol for evoking a sense of shared destiny and adaptive strategies from people. In the following year, his struggle succeeded: the first true-color photo of the illuminated face of the Earth (the terrestrial disk) was released to the public. The planetary visual consciousness was born from "the great American resource of paranoia",[55] the dominant feeling of the postmodernist epoch (why don't they show us the images? What are they hiding? We want to see it all). In the following years up until 1972, when the famous "Blue Marble" photo taken from Apollo 17 was released, a planetary visual culture was fully formed, also thanks to Brand's efforts.

Brand capitalized on his unexpectedly won bet. In 1968 he published the first issue of the *Whole Earth Catalog*, a countercultural magazine that would soon become a cornerstone of early planetary culture. Like a Surrealist magazine on drugs, the *Whole Earth Catalog* was a grimoire of concretized ideas that launched the reader into a marvelous future. Unlike Surrealist magazines, however, the Catalog—whose subtitle read "An access to tools"—was also a pseudo-encyclopedic atlas of means and tools for the modern individual to lead a good planetary life. The first issue featured book reviews on general system theory, articles on morphology and Tantra art, reportages on engineering wonders and dwelling technologies, and so on.

---

54 Stewart Brand, 'The First Whole Earth Photography', in Katz, Marsh, Thompson, 184–188 (p. 185).

55 Stewart Brand, 'Why Haven't We Seen the Whole Earth?', in Lynda R. Obst (ed.), *The Sixties* (New York: Random House, 1977), 168–179 (p. 168). "'Why haven't they made a photograph of the entire earth?' There was something wrong with 'entire'. Something wrong with 'they'. I had phrased the question "Why haven't we seen a photograph of the whole Earth yet? to follow the American enthusiasm for conspiracy, the notion that they were protecting us or keeping us from something we ought to have" (Brand, 'The First Whole Earth Photography', p. 187).

The issue's incipit was: "The Whole Earth Catalog functions as an evaluation and access device. With it, the user should know better what is worth getting and where and how to do the getting". Its purpose was the individual's "independent education". The introduction continued:

> We *are* as gods and might as well get good at it. So far, remotely done power and glory—as via government, big business, formal education, church—has succeeded to the point where gross defects obscure actual gains. In response to this dilemma and to these gains a realm of intimate, personal power is developing—power of the individual to conduct his own education, find his own inspiration, shape his own environment, and share his adventure with whoever is interested. Tools that aide this process are sought and promoted by the WHOLE EARTH CATALOG.[56]

By evoking the enhancement of human individual power against established institutions, Brand was fostering a new kind of techno-mysticism with strong evolutionary overtones. He seemed to have a practical understanding that planetary culture had to arise on the symbolical level. In a typically ideological operation, this gesture seamlessly combined naturalistic visions of the Earth from above, schemas of complex network structures suggesting the hidden osmosis between nature and technology, the psychedelic sublime, the futuristic exotic, visions on future life and evolution, the fear of war, the technological curious, etc.

The Whole Earth Catalog grew fast in readers and was published continuously up until 1972. Brand took part to the first Lindisfarne conference in 1974, where he told an anecdote about a US military officer who was appointed to investigate his activity and concluded that he represented "no serious threat to the U.S. government" (implying that he was just a Californian wackjob). In the same year, he founded the *CoEvolution Quarterly*, a magazine for the more cultivated reader which accentuated the evolutionary and populationist component. The times had changed: the hyper-modern sense of omnipotence characterizing the 1960s had largely

56 Stewart Brand (ed.), The Whole Earth Catalog. Access to Tools, Fall 1968 (p. 1).

consumed itself in social conflict, wars and stagflation. During the 1980s, Brand will be fully co-opted by the business sector.

One element that stands out in this story is the paranoid factor at the roots of planetary aesthetics. Brand explicitly declared that the initial drive to see the whole Earth was fundamentally conspiratorial and paranoid. From a psychoanalytic standpoint, the desire to see (scopophilia) derives from a schisis between the eye and the gaze: the subject desires to see the entirety of the seeable, which they feel as something that watches and threatens them (the field of the gaze), and thus to free themselves from a condition of reification and to access a (more) universal condition. The Earth is precisely a figure of the gaze: something that envelops us and cannot be seen in itself, and therefore can be perceived as both motherly and menacing. This scopic passion was historically elicited within the US culture by the "paranoid style in American politics",[57] shaped by years of McCarthyism and conservative identitarianism, as narrated by the masterpieces of American postmodernism.

Paranoia is a fundamental planetary feeling: it convinces the individual that they live in a compact world, a world that is one and whole, but also lies beyond reach. It was indeed, as McLuhan foretold, a relapse into magic thinking triggered by scientific and technological progress. The planet became simultaneously a new source of threatening opacity ("Why haven't we seen a photograph of the whole Earth yet?") and a new chance for Enlightenment. According to a known anthropological dynamic, the new-found global intimacy generated collective feelings of both suspicion and hope.[58] It is little wonder that right after the end of the post-War boom, the evolutionist component of planetary ideology quickly

57 See Richard Hofstaedter, *The Paranoid Style in American Style and Other Essays* (Cambridge, MA: Harvard University Press, 1996).

58 "Visions of prosperity and visions of conspiracy are two sides of the same coin" (Annika Rabo, 'Conspiracy Theory as Occult Cosmology in Anthropology', in Michael Butter and Peter Knight [eds.], *Routledge Handbook of Conspiracy Theories* [London and New York: Routledge, 2020], 81–93 [p. 87])

turned into survivalism; to these days, Brand sells his beliefs that extinction is an opportunity for those who are willing to take it.[59]

Stewart Brand successfully marketized the paranoid feeling into a new aesthetics: a materialist and individualist irrationalism (perfectly summarized by the motto closing the 1974 Catalog, "Stay hungry, stay foolish", later made even more famous by Steve Jobs). But a planetary aesthetics was already spreading. During his campaign, Brand had met an older guru from the Greatest Generation, Richard Buckminster Fuller, who became his spiritual and intellectual mentor. Heir of an ancient family of New England colonial aristocracy, Fuller incarnated the true utopian spirit of America, the one portrayed in his now famous dictum: "You'll never change things fighting the existing reality. To change something, build a new model which makes reality obsolete". After financial disaster and personal tragedy during the Great depression, Fuller invented the geodesic dome, which earned him a job for the US military and the fame of a visionary. His best-seller from 1969 *Operating Manual for Spaceship Earth* will later catch the *Zeitgeist* like few others.

The geodesic dome is a hemispheric structure proportionally more resistant as it grows in dimensions. As such, it perfectly embodies the scalarity of dwelling, that is the possibility of extending the human constructive act of inhabiting to the whole planet and beyond: the smaller human shelter can be structurally the same as the planet itself. Fuller's invention was the prototype of planetary design, as the reversibility between human architecture and planet lay at the roots of the new ways of understanding dwelling that took hold during the 1960s. The dome was indeed the symbolical epitome of the new cosmology. From Soleri's Dome house (a housing project in the Arizona desert)[60] to the wide deployment of Fuller's domes

59 See Stewart Brand, 'Rethinking Extinction', Aeon, April 21, 2015 (https://aeon.co/essays/we-are-not-edging-up-to-a-mass-extinction).

60 Paolo Soleri, a pupil of Franck Lloyd Wright and a Lindisfarne fellow, was an exemplary incarnation of planetary modernism. Soleri clearly identified the new architectural problem in the invention of ways to dwell on the planet. The anthill-metropolises raised issues of "complexity-miniaturization", that is how to make the household into a comfortable place opened to the technological and social cosmos (Paolo Soleri, 'The

by NASA for extraterrestrial housing units,[61] the dome represented a wordless microcosmos, a portable planet that carries the human family around the universe. Mario Merz's trademark artworks, the igloos (dating back to 1968), repeated the obsession for domes with ironic awareness of their symbolical power and anthropological implications. In general, the flourishing scene of post-war Italian art was deeply impressed by the new American myths of dwelling. In one of the exhibitions that anticipated the Arte Povera movement, *Arte abitabile* ('Habitable Art') from 1966, Piero Gilardi, Michelangelo Pistoletto and Gianni Piacentino experimented in the design of housing cells of the future after having visited the US. Ettore Sottsass, himself influenced by the time he spent in New York and Palo Alto, speculated on exotic houses and temples of the future-present where people would live as "artisan-artists" and "nomad artists" in a time "where there are no more powers, but wandering fluxes of will and public passions, which are generated within, in some way, like the molecular movements and settlements of gases and liquids".[62] The same spatial futurism suspended between criticism and fascination will manifest, for instance, in the work of Superstudio. These are just instances of how late-modernist architecture and so-called "radical

City of the Future', in Katz, Marsh, Thompson, 72–76). In line with modernism, this was conceived of as an operation of betterment through human action on reality; beyond modernism, it was also a call to rediscover frugality, spirituality, the management of scarcity and the contact with nature. On the stakes of late-modernist architecture, see Felicity D. Scott, *Architecture or Techno-utopia: Politics After Modernism* (Cambridge, MA, and London: MIT Press, 2007).

61 Fuller himself fantasized over "the universal design requirements of a scientific dwelling facility" in between the "advancement of world living standards" and the "volition to succeed / the era of 'survival'" (Richard Buckminster Fuller, 'Universal Requirements of a Dwelling Advantage', in *No More Secondhand God and Other Writings* [London and Amsterdam: Feffer & Simons, 1967], 52–73 [pp. 52–53]).

62 Ettore Sottsass, 'The Planet as a Festival', Design Quarterly, 89 (1973), 7–16 (p. 7).

design" ambiguously took on themselves the task of modeling life and the future.[63]

In 1940, Fuller—a passionate Unitarian—wrote an inelegant poem against the rising "secondhand gods", referring to all the dictators that were threatening democracy at the time. Almost three decades later, a small part of the poem was quoted in the first issue of Brand's Whole Earth Catalog. A few pages before the issue's incipit read: "We are *as* gods, might as well be good at it". Like a biblical snake, the Catalog promised to become gods through knowledge—a promise that stands at the beginning of every new culture. The omnipotence of design—that is invention, knowledge, love—was the only proof necessary of the divinity of present humanity.[64] Of course, in their revolutionary intent, designers presented their work as 'against' all established structures and doctrines, just like the early planetary ideologues. Fuller himself, rightly convinced to be part of the change as he was, made it a matter of "anticipatory design science" aimed at "the complete re-orientation of man".[65] Planetary design was unredeemably utopian.[66]

---

63 "Design does not end with the product put in production by industry but starts from that moment. It starts when it enters our homes, enters our streets, towns, skies, bodies, and souls" (Max Bill and Ettore Sottsass, 'Design and Theory', in Kathryn B. Hiesinger, George H. Marcus [eds.], *Design since 1945* [London: Thames and Hudson, 1983], 1–4 [p. 4]).

64 Not only design, but also the happening culture had a key role in the aestheticization of reality during those years. Turner recounts how Stewart Brand was influenced by the artistic community of lower Manhattan in the late 1950s and early 1960s, which included personalities like John Cage, Robert Rauschenberg and legendary performance artist Allan Kaprow. "To Brand, happenings offered a picture of a world where hierarchies had dissolved, where each moment might be as wonderful as the last, and where every person could turn her or his life into art" (Turner, p. 48). Brand would later find the same Romantic and techno-mystical spirit in the art tribe called USCO (The US Company), with which he collaborated between 1963 and 1966.

65 Richard Buckminster Fuller, 'A Comprehensive Design Anticipatory Science', in *No More Secondhand God*, 84–117 (p. 85).

66 See Lynda H. Schneekloth, 'Unredeemably Utopian: Architecture and Making/Unmaking the World', Utopian Studies, 9/1 (1998), 1–25. To

The paranoid tendencies of planetary aesthetics were best reflected in visual culture, its schizoid tendencies in design. In a way, literary culture summarized both. On the one hand, writers like Thomas Pynchon staged the fundamental dynamic of planetarization as an alienated reconstitution of the magical world.[67] With a much more glorifying intent, figures like William Irwin Thompson—himself a scholar in literature and a poet—participated in first person to the postmodernist movement; nor is it hard to link this fact with his planetary obsession. On the other hand, the literary medium too was co-opted in the creation of the future. Since Herman Kahn's *On Thermonuclear War* (1960), the founding text of futurology, writing about the future had moved beyond the boundaries of sci-fi into a mixed rhetoric combining "the expedients of science fiction, political fiction, and utopian writing", as wrote de Rosnay.[68]

In *The Macroscope*, de Rosnay rightly described it as the "scenario method". Scenario thinking and writing did not consist in planning strategies for the immediate future based on existing models, but in figuring possible long-term directions to reform the model from within. It was a method "well known to planners and futurologists and widely used" to clarify decisions and facilitate choices.[69] As taught by game theory, it is the model that dictates the possibilities; then it

---

paraphrase Manfredo Tafuri, the new tasks given to design were something besides or beyond design itself (see Manfredo Tafuri, *Architecture and Utopia: Design and Capitalist Development*, trans. by B.L. La Penta [Cambridge, Mass, and London: MIT Press], p. ix).

67 "The reader, stunned, lost, overwhelmed, battered, and mired in a crazed array of characters, plots, various conspiracies, and alliances, gropes for clues and connections, and is thus forces to try and make sense of what she is reading. The novel seems to beg to be allegorized, fit into some massive blueprint or game plan as the reader becomes her own conspirator [...]. Conspiracy becomes a lived experience, an attempted gathering of correspondences, hidden web of networks and zones" (Samuel Chase Coale, 'Conspiracy and Paranoia', in Inger H. Dalsgaard [ed.], *Thomas Pynchon in Context* [Cambridge: Cambridge University Press, 2019], 211–216 [p. 213]). See also Samuel Chase Coale, *Paradigms of Paranoia: The Culture of Conspiracy in Contemporary American Fiction* (Tuscaloosa: University of Alabama Press, 2005).

68 De Rosnay, *The Macroscope*, p. 204.

69 De Rosnay, *The Macroscope*, p. 221.

is only a matter of combinations. The penetration of this approach into political advisory unmasked the effort of naturalizing decisions through supposedly objective methods (while in fact it was only a matter of extending a limited domain of rationality beyond its own boundaries). The new cosmos was both a crushing, menacing whole that needed to be domesticized and a porous dimension invaded from all sides by the not-here and the not-now.

2.4. *Shipwreck with spectators*

In 1992, Kevin Kelly, the spokesman of Brand's *CoEvolution Quarterly*, reminisced:

> In the 1970s, after thousands of years of telling tales about the Earth's past and creation, the inhabitants of planet Earth began to tell their first story of what might happen to the planet in the future. Rapid communications of the day gave them their first comprehensive real-time view of their home. The portrait from space was enchanting—a cloudy blue marble hanging delicately in the black deep. But down on the ground the emerging tale wasn't so pretty. Reports from every quadrant of the globe said the Earth was unraveling.
>
> […] The cameras, together with reams of ground data pouring in from every country, formed a distributed mirror reflecting a picture of the whole system. The entire biosphere was becoming more transparent. The global system began to look ahead—as systems do—wanting to know what might come next, say, in the next 20 years.[70]

The first planetary consciousness was formed. Although in Kelly's recount the utopian drive does not appear as completely exhausted, the 1973–1975 recession actually marked a radical change in tone of planetary ideology. With the rise of new computing technologies (the computer as personal device) and the decline of New Communalism

70 Kevin Kelly, *Out of Control: The New Biology of Machines* (London: Fourth Estate, 1994), p. 441.

as a social movement, planetary ideology turned away from counterculture and became something else.[71]

After the conclusion of Brand's second editorial adventure, the legacy of planetary futurism was taken up by the magazine *Wired* starting 1993. By then, the Cold War Left had given way to that which Barbrook and Cameron called the "Californian ideology."[72] The "children of McLuhan" had become the backbone of American neoliberalism. The new "techno-élite" was tasked by figures like Hans Moravec, Marvin Minsky and Ray Kurzweil to free pure intelligence into the universe[73] and bring about the obsolescence of the body. Cyberspace and outer space alike represented the radical de-materialization of the planetary utopia, or rather its fulfillment as a cosmological vector of illocalization. The American transhumanists devoted themselves to perfection that sort of materialist spiritualism (or spiritualist machinism, to paraphrase the title Ray Kurzweil's book from 1999) which had been invented by Brand, Fuller and Thompson in the previous decades.

Kevin Kelly himself, as editor of *Wired*, resumed their predictions on the emergence of a global mind within a network culture at the dawn of the public Internet. Kelly argued that, due to the complexity of this new "union of computer and nature" "governed by an invisible hand of its own", "we humans will be unconscious of what the global mind ponders, [...] because the design of a mind does not allow the parts to understand the whole. The particular thoughts of the global mind—and its subsequent actions—will be out of our control and beyond our understanding. Thus network and economics

---

71 See Turner, especially chapter 4; see also Erik Davis, *Techgnosis: Myth, Magic and Mysticism in the Age of Information* (New York: Crown Publishers, 2004).

72 Richard Barbrook and Andy Cameron, 'The Californian Ideology', in Richard Barbrook, *The Internet Revolution: From Dot-com Capitalism to Cybernetic Communism* (Amsterdam: Institute of Network Culture, 2015), 12–27.

73 See Hans Moravec, *Mind Children: The Future of Robots and Human Intelligence* (Cambridge, MA: Harvard University Press, 1988), p. 116. The notion of "techno-élite" appears in the work of Ray Kurzweil. See André Gorz, *The Immaterial: Knowledge, Value Capital*, trans. by C. Turner (London and New York: Seagull Books, 2010), chapter 4.

will breed a new spiritualism".[74] Technology had finally naturalized itself; the social body had been successfully reduced to a swarm condition. It was the social order dreamt by cybernetics: "Control without authority".[75]

The end of the Cold War set the conditions for an internal phase-shift of System cosmology: planetarization properly so called. Herein lies a bifurcation that I described at the end of chapter 1 between pro- and anti-system cosmology—where 'system', let us recall, is the framework provided by systems science. On one side, pro-system cosmologists enshrined the planetary as the grand narrative of the present age by re-launching a form of modernist rationalism. On the other, anti-system cosmologists inherited the anti-modernist irrationalism of early planetary ideology, either in an accelerationist or ecologist sense. Still today planetary ideology keeps together these many souls, often indistinguishably.

The accelerationists took on the whole-Earth techno-mysticism and brought it to a new level of expression. The most symptomatic case is the Cybernetic Culture Research Unit (CCRU), an esotericist cluster of thinkers gathered at the University of Warwick. Accelerationism was an outgrowth of British cultural theory born at the height of globalization.[76] After cyberfeminist philosopher Sadie Plant abandoned the group and Nick Land took over in 1997, the CCRU turned into an explosive mix of cybernetic communism of the net and posthumanist techno-mysticism that got rid of the irenic tone prevailing in systems theory to assume a markedly aggressive and apocalyptic style.

Interestingly enough, scenario thinking was identified within the CCRU as a device of manipulation of the future intrinsic to capitalism. Capitalist prophecies, they argued, tend to self-fulfill. "California is waiting for the Big One", wrote Plant and Land back in 1994.[77] The illusion of Landian CCRU was to harness this

74 Kelly, p. 202.
75 Kelly, p. 26.
76 See Edmund Berger, *Accelerazione. Correnti utopiche da Dada alla CCRU* (Roma: Nero, 2021), chapter 13.
77 Sadie Plant and Nick Land, *Cyberpositive*, in Robin Mackay and Armen Avanessian (eds.), *#Accelerate#: The Accelerationist Reader* (Falmouth:

prodigious power of future manipulation by separating the market from capitalism and any other social system to free its a-human forward-driving force. The subject of history, they posited, must adjust to this catastrophic vector; it must become a non-subject through cybernetic de-individuation. CCRU's anarcho-capitalism nurtured the fundamental fantasy that capitalism will eventually free itself from itself, and that liberation from the systems of oppression has nothing to do with will and freedom: "Revolution is not a duty, but surrender".[78] This paradoxical non-tactic manifested in full in the group's writings, which mimick the psychedelic iridescence of a Baudrillardian world.

From a cosmological standpoint, the CCRU and its epigones formulated a sort of nihilist cosmism echoing some aspects of Deleuze and Guattari's geophilosophy. Elaborating on the idea of "geotrauma", authors such as François Laruelle, Ray Brassier, Eugene Thacker, Ben Woodard and Reza Negarestani argued in different ways that the planet is crossed by a fundamental death drive, or a vector of abstraction, that manifests its belonging to the universal folly. Contrary to the anthropocentric and phenomenological picture of the Earth as a motherly cradle and a stable ground for human experience, a "georealism" acknowledges the planet as a potency of

---

Urbanomic, 2014), 303–313 (p. 305).

78 Nick Land, 'Making it with Death: Remarks on Thanatos and Desiring-Production', in *Fanged Noumena. Collected Writings 1987–2007*, ed. by Robert Mackay and Ray Brassier (Falmouth: Urbanomic, 2012), 261–288 (p. 287). "The identity of revolutionary and anti-fascist politics lies in resisting capital's molar projection of its death. All the supposedly alien sources of disorder which capital represents as the exteriority of its end, such as working class agitation, feminism, drugs, racial migration, and the disintegration of the family, are as essential to its own development as the attributes of a substance. The revolutionary task is not to establish a bigger, more authentic, more ascetic exteriority, but to unpack the neurotic refusal mechanisms that separate capital from its own madness, luring it into the liquidation of its own fall-back positions, and coaxing it into investing at the deterritorialized fringe that would otherwise fall subject to fascist persecution. Schizo-politics is the coercion of capital into immanent coexistence with its undoing" (pp. 276–277). This means aligning oneself with the most advanced formations of capitalism—a strategy that so far has always and obviously fueled the restructuring of capitalism rather than overcoming it.

destitution. It is easy to see how the destructive critique of humanism turns into the naturalization of capitalist abstraction, one of the less anthropocentric operations in the history of human civilization. For these thinkers as well as for the latest technoscientific and mythical vectors of late-liberalism, the planet is only a platform to be left behind.

The epigones of CCRU, concerned and shocked by Land's reactionary swing, have strenuously tried to reconnect accelerationism with rationalism. Accelerationist cosmology understands itself according to the "Copernican imperative"[79] of overthrowing the humanist morals and disrupting the passion for the obvious through scientific realism. Here lies an important zone of osmosis with the planetary engineers. Yet, in both cases, a planetary rationalism leads to blatantly irrational conclusions: either, the Prometheanism of planetary-scale technological projects, apparently ignorant of any principle of ecology, or a mystical negation of praxis and a sanctification of pure speculative reason, a cadaveric escapism insufficiently justified by the polemics with modern anthropocentrism.

The fundamental illusion of early planetary utopianism, reverberating up to its last ideological ramifications, was believing that technology could reform the social, economic and political system, while it was this very system that created the technology.[80] As the political myth of globalization was gradually dispelled, the beliefs that the planetary system can self-organize in the absence

79 See Demien Veal (ed.), *Collapse*, vol. V, *The Copernican Imperative* (Falmouth: Urbanomic, 2009).

80 See Maurizio Lazzarato, 'On the Californian Utopia/Ideology', in Diederich Diederichsen, Anselm Franke (eds.), *The Whole Earth: California and the Disappearance of the Outside* (Berlin: Sternberg Press, 2013), 166–168. "During the financial crisis", explains Lazzarato, "machines failed to provide for any measure of self-regulation [...]. On the contrary, their automation often amplified the disequilibrium. The machines did not demonstrate any 'autonomy', because they always operated within a network of which they were merely constituent parts, intentionally open and connected to 'otherness' and 'externalities', of which the most important were proved once again to be the workings and valuation of 'capital'" (p. 167).

of authority, that we find ourselves on an evolutionary slide and that we must therefore create a new universal culture matching the techno-scientific acceleration became more and more grotesque. Planetary ideology, however, did not disappear: it rather came out of its embryonic phase and manifested in full. The long migration of planetary utopianism from Left to Right and the conversion of utopianism into apocalyptic extinctionism are part of this trajectory.

Today the heirs of accelerationists and transhumanists, reunited into the ideological bundle of TESCREAL (the acronym for Transhumanism, Extropianism, Singularitarianism, Cosmism, Rationalists, Effective Altruism, Longtermism),[81] explicitly advocate for a general escapism and an ideology of illocalization. Like the early planetary thinkers, the TESCREALists—sources of inspiration for Elon Musk, Jeff Bezos and Richard Branson—think in terms of deep space (ours must be seen as one planet among others) and deep time (our values must be calibrated on transgenerational and ultimately evolutionary challenges). Since planetary consciousness (or "planetary sapience"[82]) has been accomplished, 'we'—embodiment of the planetary self-awareness—should now promote planetary-scale actions, on this or other planets. American philosopher Benjamin Bratton explicitly conceptualizes planetary design as "terraforming", originally a word for modifying extraterrestrial planets into being earth-like. Bratton's proposal is impressive in its clarity: a "pro-universalist, anti-anti-totality", "anti-anti-leviathan" and "pro-egalitarian" program for urban inhabitation.[83] No mention is obviously made as to the material and social dimension of this

81 See Timnit Gebru and Émile P. Torres, 'The TESCREAL Bundle: Eugenics and the Promise of Utopia through Artificial Intelligence', First Monday, 29/4 (2024), https://doi.org/10.5210/fm.v29i4.13636, where the acronym was first coined.

82 Benjamin Bratton, 'Planetary Sapience', in Gilman, 35–57. Elsewhere "planetary sapience" is defined as "a technologically enabled self-understanding of the planet and its deeply systemic interconnectedness" (Jonathan S. Blake and Nils Gilman, *Children of a Modest Star: Planetary Thinking for an Age of Crises* [Stanford: Stanford University Press, 2024], pp. 6–7).

83 Benjamin Bratton, *The Terraforming* (Moscow: Strelka Press, 2019), p. 8.

operation. If the Earth is a spaceship, these thinkers desire to be the shipwreck's spectators.

Today this ideological cluster is more transversal and opaquer than it may look, especially in its aesthetic manifestations. "Post-industrial society", the social paradigm of planetary ideology,[84] has already given birth to its ambiguous poets. Let us take Neri Oxman, whose ecomodernist design aims at going beyond the industrial premises rooted in Western culture regarded as the source of all evil. "The world of design", she argues, "has been dominated since the Industrial Revolution by the rigors of manufacturing and mass production. Assembly lines have dictated a world made of standard parts framing the imagination of designers and builders who have been taught to think about their design objects and systems in terms of assemblies of parts with distinct functions".[85] Instead, today's design and manufacturing technologies can program the growth of matter at the molecular level and adapt it to new behaviors; the human body, for instance, can be endowed with devices that allow it to survive on other planets.[86] Speculative design and ecodesign are now powerful allies in guiding humanity towards a new Eden after having atoned for the sins of modernity. Their objective is comfort in the present through the promise of happiness in the future (precisely the task that once belonged to art).

Another example of ambiguous contemporary poetics is the neo-cosmist artist and founder of e-flux Anton Vidokle, whose work re-enchants the old Cosmist dream of colonizing the universe and achieve "immortality and resurrection for all"[87] through science and technology. There is a productive schisis internal to Vidokle's works: on the content level, the hyperbolic dogmas of Cosmism are emphatically proclaimed like an ancient dream in which one can still

---

84 See note 46.

85 Neri Oxman, 'Editorial. Material Ecology', Computer-Aided Design, 60 (2015), 1–2 (p. 1).

86 See Neri Oxman, 'Dermi-Domus: A Grown Wardrobe for Bodies and Buildings', Architectural Design, 87/6 (2017), 16–25, and https://deskriptiv.com/wanderers.

87 Anton Vidokle, 'Immortality and Resurrection for All!', in *Citizens of the Cosmos* (Berlin: Sternberg Press, 2024), 165–183.

believe; on the formal level, the power of these ideas is unleashed on the contemporary spectator through constant contrasts between sublime and bathetic images; finally, on the enunciative level, it remains unclear if Vidokle is carrying on his discourse as a member of a present, past or future collectivity. The spectator is left with a universally manageable and easily capturable utopia of the form. Unfortunately, it is not enough to utter the word 'communism' to craft a progressive mythology; planetarity, after all, has always been deeply egalitarian.

The same ultra-modernist belief animates the sublime aesthetics of the aerial visions of the Earth (from the *Whole Earth Catalog* to the omnipresent iconography of the Anthropocene[88] and the idea of planetary mediality[89]) as well as the ecomodernist beautiful, the planetary neo-epic and the posthuman weird (for instance in the works of Pierre Huyghe, Hernan Diaz Alonso or Alisa Andrasek). These poetics contribute to enchanting and accelerating the distortions of the present; and in this sense, their sublimating effect is a manipulation of transcendence that aims to bring the present beyond itself. To be sure, the complexity and opacity of these cultural products prevent us to reduce them to simple mirrors of their time. But the degree of penetration of the social imagery is itself an aesthetic value according to which one can pass judgement on an artwork's idleness.

There are indeed positions embracing the planetary view without explicitly or implicitly endorsing the productive system that achieved it. Patricia Reed, for example, attempted to outline a

88 The Anthropocene sublime is a widespread category that usually hinges on the work of photographers such as Edward Burtynksy and Jennifer Baichwal (see e.g., Andrea Borsari, 'On the Aesthetics of the Anthropocene: The Sublime and Beyond – Other Concepts and Forms of Visualization', European Journal of Creative Practices in Cities and Landscapes, 5/2 [2022], 242–258). Many cases—like *Slow Violence* by Studio Lemercier (https://joanielemercier.com/slowviolence/), just to name one—demonstrate that this poetics is the subject of a wide critical reappropriation.

89 See e.g., Tiago de Luca, *Planetary Cinema: Film, Media and the Earth* (Amsterdam: Amsterdam University Press, 2022).

new understanding of human care in the "big-world continuum".[90] Contemporary artists such as the Otolith Group propose a planetary futurism well situated within the tradition of the oppressed.[91] "Alternative modalities of sublimation"[92] can indeed be produced that criticize the mythologies of the present; abstraction can be a way to reappropriate the powers of design after usefully dispelling the fetishes of concreteness, individual genius, naturality, a-conceptuality and experiential immediacy.[93] Yet too often do the advocators of abstraction "retrain the work of the 'political' away from a direct confrontation with or acceleration of Capitalism *as the scope of the problem as such*".[94]

From this angle, the planet appears as an anti-anthropocentric event torn between extinctionist and Promethean delusions. The main difference between these two osmotic aspects is practical: disembodied reason can either inspire a new faith in the human action or call for a final surrender of the Anthropos. Abstraction can endow 'us'—humanity as an abstract planetary subject—with either more agency or less. The ultimate ground of discrimination between these two souls of planetary ideology is therefore the sphere of action and decision. Both accelerationists and geoengineers, however, remain on the side of abstract fungibility, either to deny or to magnify it.

---

90 Patricia Reed, 'What is Care at Planetary Dimensions?', lecture given at Floating University, Berlin, August 6, 2019 (https://laboriacuboniks.net/texts-interviews/what-is-care-at-planetary-dimensions/).

91 See T.J. Demos, *Radical Futurisms: Ecologies of Collapse, Chronopolitics, and Justice-to-Come* (Berlin: Sternberg Press, 2023), especially chapters 1 and 2.

92 Mark Fisher, 'Left Hyperstition 2: Be Unrealistic, Change What's Possible', K-punk, November 29, 2005 (https://k-punk.org/left-hyperstition-2-be-unrealistic-change-whats-possible/).

93 Nick Srnicek, 'Accelerationism – Epistemic, Economic, Political', in Robin Mackay, Luke Pendrell and James Trafford (eds.), *Speculative Aesthetics* (Falmouth: Urbanomic, 2015), 48–53 (p. 52).

94 Benjamin Bratton, 'Some Trace Effects of the Post-Anthropocene: On Accelerationist Geopolitical Aesthetics', e-flux, 46 (2013) (https://www.e-flux.com/journal/46/60076/some-trace-effects-of-the-post-anthropocene-on-accelerationist-geopolitical-aesthetics).

3.
THE EARTH AS PATHOS

3.1. *The apolitical Gaia*

> In practice, we are all counter-revolutionaries.
>
> Bruno Latour, *Facing Gaia*

The other cosmology stemming from early planetary ideology was Gaia cosmology. Like the accelerationists, the Gaians refuse to see the Earth as an appropriable object or a system whose behavior is aimed at its own stability. Yet they carry on symmetrically opposed values: slowing down instead of accelerating, concretization instead of abstraction, ecology instead of technology, Epimetheanism instead of Prometheanism. "Dream no longer, mortals! You won't escape into space. You have no dwelling place but this one, this narrow planet".[1] The Gaians express a planetary universalism in direct competition with that of the accelerationists—one that has nothing to do with the 'small is beautiful' spiel that is often attributed to them.[2]

Let us go back to the Lindisfarne Association. In 1981, the annual conference of Thompson's creature was held in San Francisco and hosted some of the most relevant scientists and thinkers of the time: besides the longtime fellow Bateson, Francisco Varela, Humberto

---

1 Latour, *Facing Gaia*, p. 81.

2 On the opposition between accelerationists and Gaians, see Harrison Fluss and Landon Frim, *Prometheus and Gaia: Technology, Ecology and Anti-Humanism* (London and New York: Anthem Press, 2022). This book effectively demonstrates the differences and the similarities between the two, although it makes no difference between the Prometheanism of planetary engineerism and that of accelerationism.

Maturana, James Lovelock, Lynn Margulis, Henri Atlan—the royalty of systems science. A crucial event had breathed new life into Lindisfarne's task of creating a planetary culture: the publication of James Lovelock's *Gaia: A New Look at Life on Earth* in 1979. Lovelock—with the fundamental collaboration of Lynn Margulis—had done more than provide a systemic explanation of the Earth as an autonomous entity: he had given shape to the epoch's cosmology and introduced a new mythological persona. Gaia, the planet, was now a being comparable to earthly organisms and to every other phenomenon of order in the universe, that is a being complex enough to regulate itself in the exchange with its environment.

The political implications drawn by the anti-Reagan intellectual Left (as Thompson defined Lindisfarne's official position) were clear. Humanity found itself in a phase of terror and international chaos which would soon bring forth the realization of the planetary utopia. With the Reagan presidency, globalization was showing its ruthless, terroristic side; but biology taught that catastrophe is just transition and what is evil today can be good tomorrow. Once aggressively utopian, the planetary discourse was fading into an invitation to surrender to planetary immanence. "Part of the Gaian politics for the nineties, argued Thompson, is [...] to realize that we are all organelles within a planetary cell, and that it is a dangerous illusion to think that any nation-state can make it on its own, militarily or economically".[3]

"Gaia politics", as Thompson called it, was based on the idea that all political actors were more or less the same in their struggle to exist in spite of the others, and that ideology condemned them to chaos and conflict. But systems science had shown that such oppositions were tied to a limited viewpoint. The sciences of nature pointed towards the relativity of human values and the overcoming of all ideologies. "One cannot say that the ocean is right and the continent is wrong

3 William Irwin Thompson, 'Gaia and the Politics of Life. A Program for the Nineties?', in William Irwin Thompson (ed.), *Gaia, a Way of Knowing: Political Implications of the New Biology* (Great Barrington, MA: Inner Traditions/Lindisfarne Press, 1987), 167–214 (p. 193).

in a Gaian view of planetary process".[4] "If one is an aerobic cell, oxygen is evil; if one is a hominid, human tools are evil; if one is a hunter or a nomad, agriculture is evil; if one is a farmer, industrial cities are evil; and now if one is a civilized, literature human being, a noetic ecosystem (or Chardin's Noosphere) is evil".[5] Consequently, "the movement from archaic industrial modes of thought into a new planetary culture is characterized by a movement from ideology to an ecology of consciousness".[6]

The task for the following decade (the 1990s), prophesized Thompson, would be to generate awareness around the much-anticipated advent of post-industrial society and the unacknowledged reality of the planetary age.[7] Gaia politics would be post-ideological and post-hierarchical, because Gaia reunites all standpoints under an immanent principle of totality and as such does not cancel the differences, but let them blossom. Since the planet does not subsume the many orders of its beings, but is rather shaped by their interaction and emerges as their hyper-environment, the Gaia polity would not be a new ideology but embrace all ideologies (which, it is implied, would disappear as such and become something else entirely). All this discourse is grounded on a passage from the moral and political indifference of ontological relations to the moral and political indifference of political relations—a passage that the social sciences were called to infer from the new biology. Gaia politics, then, as an apolitical politics.

With the running aground of the 1960s utopias, the belief in the possibility to drive human evolution through knowledge and technology had been gradually replaced by the acceptance of the planet's intrinsic indeterminism. The faith in technology as a motor

4 William Irwin Thompson, 'Introduction', in Thompson, *Gaia, a Way of Knowing*, 11–34 (p. 27).
5 Thompson, 'Gaia and the Politics of Life', p. 181. Note the polemic tone towards the old intellectual class, that of the "civilized, literate human being", to be supplanted (or rather already supplanted) by a new class, the Gaia politicians.
6 Thompson, 'Gaia and the Politics of Life', p. 181.
7 William Irwin Thompson, 'It's Already Begun: The Planetary Age is an Unacknowledged Reality', The New Story, Winter 1985/86, 26.

of social and anthropological change had given way to the faith in the nature described by systems science. To overcome the nature/culture divide now meant to let go of the old anthropocentric pretensions of controlling and manipulating nature and to invent a new true-to-nature culture. "The 'Man' of the historical set of Culture-Nature will come to his end in a new irrational world of angels and devils, elementals and cyborgs", wrote Thompson programmatically.[8] Gaia politics would have to be post-humanist and post-rationalist.

After the triumph of American capitalism in 1989, however, these visions had to adjust to the fact that a new ecumenic culture had not in fact emerged from the chaos and one of the two planetary contenders had simply prevailed. Those who wanted to carry on the Gaian utopia were faced with two choices: either becoming minstrels of the winner and confessing that planetarization had always been the same as American globalization or admitting that the dream was over and raising the bet on different grounds. The accelerationists chose a half-measured solution: America won because it let itself be possessed by the demon of history, the same that would bring the American empire itself to its demise. In all its delusion, this standpoint had the merit of grasping a simple fact: from now on, every form of resistance to the system would be *internal to the system itself.* The eschaton had come, but it was not as expected; now the only possible way forward was to amplify its intrinsic contradictions.

The Gaian strain took on an analogous but opposite approach. Bruno Latour will later reminisce that the Reagan era was the moment when globalization showed its true colors.[9] After 1989 it became definitely clear that a certain idea of planetarity was coming to an end: the idea that the planet was a goal to be attained, a finality of history. From then on, continues Latour, the discourse on the planet became the narrative of the few who dream to leave *this* planet behind. The Gaians, instead, aimed to invent another planet by transforming Lovelock's lesson. "Contrary to all the other

8 Thompson, 'Introduction', p. 27.

9 Bruno Latour and Dipesh Chakrabarty, 'Conflicts of Planetary Proportions: A Conversation', Journal of the Philosophy of History, 14 (2020), 419–454 (p. 427).

planets before, in this one the Earth has become an agent and no longer the backdrop of human agency".[10] Gaia was the true subject of history; even more, the true subaltern of history, the remainder of modernity: not a divinity of contingence and harmony, but an outlaw of knowledge.

Lovelock's description of Gaia as a self-stabilizing and self-regulating system was corrected into the idea of Earth as "the anti-system"[11] and the "counter-power".[12] The Gaian planet is non-deterministic, non-linear, potentially catastrophic: an epistemic "intruder", a conflictual and relatively alien transcendence[13] that inspires even "a sense of panic in the face of a loss of control".[14] Information is genesis, not codification. Differently from accelerationism, however, this kind of anti-systemic approach prioritized concretization over abstraction: Gaia forces us, the abstracting beings, to be here and now, to situate ourselves on this planet. Our ideas (including our political and religious beliefs) are mere fantasies in front of the Earth. Mind is put back in the body and the body in the earthly multiplicity. The principle of immanence already described by Thompson is therefore refined into a full-fledged agential monism—as in the famous Actor-Network theory developed by Latour from the 1980s onwards.

Latour is a paradigmatic example of how systems theory gave rise to a new planetary political theology. His arguments, like those of the other authors I mention in passing, are well known and I will not reconstruct them here.[15] Especially in his 2013 Gifford Lectures

---

10 Latour and Chakrabarty, p. 432.

11 Latour, *Facing Gaia*, p. 87.

12 Latour and Chakrabarty, p. 432.

13 Stengers, p. 47. "The intrusion of this type of transcendence, which I am calling Gaia, makes a major unknown, *which is here to stay*, exist at the heart of our lives. This is perhaps what is most difficult to conceptualize: no future can be foreseen in which she will give back to us the liberty of ignoring her".

14 Déborah Danowski and Eduardo Viveiros de Castro, *The Ends of the World* (Cambridge, UK, Malden, USA: Polity Press, 2016), p. 80.

15 For other critical outlooks on Latour besides those directly quoted in what follows, see Matthew C. Watson, 'Cosmopolitics and the Subaltern: Problematizing Latour's Idea of the Commons', Theory, Culture &

and in *Down to Earth: Politics in the New Climate Regime* from 2018, carrying on an operation begun in *Politics of Nature* (1999), Latour takes charge of drawing the boundaries of Gaian cosmology. Gaia is described as the multiplication of "Providences", "as many Providences as there are organisms on Earth": a "muddle" of agencies in itself capable of dismantling all modern convictions of humanism and modernism.[16] Here we find again the tropes of planetary ideology, only in a more refined guise: the Hobbesian presupposition that a new synthetic perspective is the only antidote to chaos; the postulated need of a horizontal, neo-pagan form of transcendence revolving around the natural and social sciences that promote indeterminism, geneticism and complexity; a systematic de-historicization and intellectualization of social and political dynamics in a post-ideological sense;[17] a deeply interiorized re-westernizing impulse lurking behind the infinite multiplication of standpoints and rationalities.[18]

Latour speaks as an intellectual concerned with the decline of the West and proposes plans "in preparation for the times when we shall no longer be in a position of strength and when the others will be the ones purporting to 'modernize'—but in the old way and, as it were, without us".[19] The multiplication of rationalities is the necessary acknowledgement for the wounded West to propose a new social pact *in extremis*: a new parliamentarist proceduralism that allows it

Society, 28/3 (2011), 55–79; Gunnar Skirbekk, 'Bruno Latour's Anthropology of the Moderns: A Reply to Maniglier', Radical Philosophy, 189 (2015), 45–47; Barbara Hernstein-Smith, 'Bruno Latour's Climate Evangelism', New Formations, 107–108 (2022), 230–235.

16 Latour, *Facing Gaia*, p. 100.

17 The aim is admittedly that of producing a superordinate "alternative to the Left/Right dichotomy" (Bruno Latour, *Down to Earth: Politics in the New Climatic Regime*, trans. by Catherine Porter [Cambridge, UK and Madford, MA: Polity Press, 2018], p. 52).

18 On the latter point in particular see Federico Luisetti, 'Demons of the Anthropocene: Facing Bruno Latour's Gaia', Philosophy Kitchen, 5/3 (2016), 175–186; and Id., 'Decolonizing Gaia, or, Why the Savages Shall Fear Bruno Latour's Animism', Azimuth, 9 (2017), 61–70.

19 Bruno Latour, *An Inquiry into the Modes of Existence* (Cambridge, MA: Cambridge University Press, 2013), p. 16.

to retain a seat at the table and pass on the old wisdom of modernity. All this, however, remains hidden behind the universalist naturalism of Gaia. To "repoliticize" ecology is to "reterrestrialize" our existence,[20] that is to dissolve politics into ontological immanence and its corresponding form of knowledge. This gesture, as Peter Osbòrne rightly observes, requires the evacuation of "any social content that cannot be reduced to the abstract spatial difference between the local and the global".[21]

Latour is careful to untie himself from the most regressive aspects of planetary ideology. But globalization, he says, is too good a word to let go.[22] One must simply distinguish between imperialistic globalization—the one that seeks to reduce the planet to the interests of a few—and a globalization that is simply a widening of our world ("*multiplying* viewpoints, *registering* a greater number of varieties, *taking into account* a larger number of beings, cultures, phenomena, organisms, and people"). With the necessary epistemic adjustments, the universalist aspirations of the West may continue. Likewise, there are also a good and a bad localism: reactionary localism collapses the world into the soil, while progressive localism goes deep into the infinite richness of terrestrial reality. By recycling old dichotomies, the insistence on a new knowledge and a new culture as a cure-all for the distortions of the present (summarily identified with modernization) continues to hunt the planetary discourse in the absence of any reference to concrete political or social conflict.

Similar cases could be made against other cornerstones of Gaia cosmology, like Donna Haraway's romantic notion of sympoiesis. Even more telling, however, is Gaia aesthetics. Gaia aesthetics is an aesthetic of concreteness, embodiment and worldliness that exalts the values of ontological imagination and compassion already suggested by Thompson. It is an aesthetics *of* the Earth as a texture

---

20 Latour, *Facing Gaia*, p. 223.

21 Peter Osborne, 'The Planet as Political Subject?', New Left Review, 145 (2024), https://newleftreview.org/issues/ii145/articles/peter-osborne-the-planet-as-political-subject. The call for a political ecology here is little more than a professedly neoliberal "generalization of the mode of life of the rural French petit-bourgeoisie" still dreaming of a just globalization.

22 Latour, *Down to Earth*, p. 16.

of affects and agencies, to which human experience—as described for instance by the 4E cluster in cognitive sciences (embodied, embedded, extended, enacted)—fundamentally belongs. In this sense, every terrestrial entity from the stone to the mind can be seen as an actor and a medium. This entails the possibility of a collaborative construction of the gaze from the middle of the relation, as in the case of "non-human photography", consisting in images emerging spontaneously from biochemical or machinic responses; or ecological dramaturgy, which extends the stage to other species and earthly entities; or again all "arts of living on a damaged planet" regarded in their aesthetic value.[23]

An ecological turn can be observed across all artistic fields and a great number of cultural phenomena. In general, Gaia aesthetics takes charge of naturalizing the material transcendence of the planet through concrete encounters in which there is no spectator, only performers. In line with the tenets of Gaia cosmology, it also tends to erase the ontological separation between art (regarded as a distinctively human activity) and reality through the naturalization of alterity.

3.2. *Postcolonial planetarity and decolonial worlds*

In continuity with the materialist theology of the Gaians are the attempts to recompose a cosmology in a post- and decolonial perspective. Symptomatically speaking as a counterpoint to Latour, Dipesh Chakrabarty recalls:

> 1989 was not just about the significance of the fall of the Berlin Wall for Western democracies. One should not forget the Tiananmen Square protests of the same year. Nor was the story exclusively about

---

23 See respectively Joanna Zylinska, *Nonhuman Photography* (Cambridge, MA: MIT Press, 2017); Lisa Woynarski, *Ecodramaturgies: Theatre, Performance and Climate Change* (Cham: Palgrave MacMillan, 2020); Anna L. Tsing, Nils Bubandt, Elaine Gan, Heather Anne Swanson (eds.), *Arts of Living on a Damaged Planet: Ghosts and Monsters of the Anthropocene* (Minneapolis and London: University of Minnesota Press, 2017).

> the "moral clarity" of the West—the colonized and the Western critics of empires had always known how flawed and misleading (for the West itself) that clarity was, since it accompanied and justified every piece of colonial aggression. The late 1980s and the 1990s remain important, also, for what happened in the non-Western world and in postcolonial thinking.[24]

The planet of decolonial and postcolonial thinkers has never been one; yet they have been often tempted by cosmological universalization—a truer, more universal one, as in Aimé Césaire's "true humanism [...] made to measure the world" (a "Europeanization" of the world without "Europe *occupation*").[25]

Édouard Glissant, for instance, imagined a "Whole World" emerging as an incalculable effect of colonization: a "Chaos-World" endowed with a paradoxical wholeness consisting in a heterogeneous and dynamic mixture of groups.[26] Decolonization, here, is not just an act of resistance, but also one of recomposition. In order to distance himself from the "strategic essentialism" of the Negritude movement, Glissant seemingly ended up advocating for a complacent postmodernist mosaic that gets along perfectly with neoliberal immanentism.

In the same year (1997), Gayatri Spivak uttered her famous imperative to re-imagine the planet, in which occasion she coined the word "planetarity" by setting it apart from the planetary, the Earth, the World and the globe of globalization. While the planetary is a physical totality that confines its own alterity into an abstract region which will never speak back (outer space), planetarity—she argued—is the internal alterity of globalization represented by human difference and inhabitation. It is here, in human plurality, that resides the responsibility towards each other and the planet. Ecology itself cannot be trusted in the hands of the "white mythology of

24 Latour and Chakrabarty, p. 451.

25 Aimé Césaire, *Discourse on Colonialism*, trans. by J. Pinkham (New York: Monthly Review Press, 1972), p. 56.

26 Édouard Glissant, *Treatise on the Whole-World*, trans. by C. Britton (Liverpool: Liverpool University Press, 2020), p. 12.

post-rational science",[27] but must rather rely on the human practices left behind by modern Reason. Thus Spivak acknowledges "post-rational science" (i.e., the whole systems science milieu) as a "white mythology", but does not leave it out of planetary thinking. The same happens with "culture", precisely for its virtuous claim "to a pattern of behavior beyond reason alone".[28] In its polemics with an abstract Reason, its accommodating inclusivity and its diplomatic appeal to universal human values (the notion of imperative is explicitly Kantian), Spivak's planetarity appears as a way to fight for a place in the post-1989 world armed with good intentions.

Regardless of how all this may sound to the European reader, one must ask what kind of universalism is the one vindicated by non-European subjectivities and collectivities. Even the words used should be heard differently than when used by others. My criticisms is limited to their usability and the possibility to extend them beyond their context regardless of their relative strategic function. From this perspective, it is immediately visible how, contrary to the Gaians, the advocators of postcolonial planetarity rely on the notion of an irreducibly human experience. The subject is humanity once again; and the planetary question "seems to be for humans alone".[29] But how is this humanity (this 'we') characterized? Here postcolonial theory fluctuates between a humanistic notion of meaningful experience, the coinage of new universals[30] and a vague concept of human commonality.[31] Its most relevant contribution seems to lie more in

27 Gayatri Chakravorty Spivak, 'Imperative to Re-Imagine the Planet', in *An Aesthetic Education in the Era of Globalization* (Harvard: Harvard University Press, 2012), 335–350 (p. 339).

28 Spivak, p. 341.

29 Dipesh Chakrabarty, *One Planet, Many Worlds: The Climate Parallax* (Waltham, MA: Brandeis University Press, 2023), p. 12.

30 To the ones already mentioned, one could add for instance Chakrabarty's "Dalit body" (Dipesh Chakrabarty, *The Climate of History in a Planetary Age* [Chicago and London: University of Chicago Press, 2021]).

31 Besides the already mentioned contribution by Spivak, see for instance Achille Mbembe, *Out of the Dark Night: Essays on Decolonization* (New York: Columbia University Press, 2021), p. 41. "It is about radical sharing and universal inclusion. It is about humankind's implication in a common that includes nonhumans, which is the proper name for democracy". For

the strategic effort of imagining *another* planet starting from the forced planetarization of places (as in Mbembe's "planetary turn of the African predicament"[32]) than in the actual solutions it proposes.

At first glance, the decolonial perspective seems more consistent in its political critique of globalization. Modernity is typically depicted by decolonial thinkers as a series of operations of "noming" (imposition of one *nomos*) by hand of the European empire onto the many pre-existent *nomoi*; an operation echoed today by the many attempts at "re-noming", that is "the appropriation and expropriation of land by international corporations with the cooperation of nation states"[33] in the former colonial countries. Strategies inspired to Third World internationalism (such as Amin's "delinking"[34]) and to indigenous resistance (such as Mignolo's "de-noming") as well as the reclaiming of the planetary commons make decolonial thinking one of the most significant alternatives to planetary totalization.

Yet, the decolonial option usually fails to produce an alternative political cosmology and condemns itself to localism. Anna Tsing paradigmatically argued that any attempt to scale up praxis and existence is essentially compliant with the capitalist logic (economies of scale, expansion and equivalence subordinated

Mbembe, "decolonization itself", "as an act of rebellion turned into an act of refoundation", opens up to postcolonial thinking as a planetary thinking, that is as a "thinking on a world scale: thought capable of taking into account the common history that colonization has made possible" (p. 44). Elsewhere, he writes: "The last utopia consists in coming back to the Earth, the last name of a *we* that would include human beings as well as objects, viruses, plants, animals, oceans, machines [...]. The great obstacle to the idea of a 'we' is, however, racism, the ultimate neurosis of separation" (Achille Mbembe, *The Earthly Community: Reflections on the Last Utopia*, trans. by S. Corcoran [Rotterdam: V2_, 2022], p. 125).

32 Mbembe, *Out of the Dark Night*, p. 10.

33 Walter D. Mignolo, 'Foreword. Anomie, Resurgences, and De-Noming', in Federico Luisetti, John Pickles and Wilson Kaiser (eds.), *The Anomie of the Earth. Philosophy, Politics, and Autonomy in Europe and the Americas* (Durham and London: Duke University Press, 2015), vii–xvi (p. xii).

34 See Samir Amin, 'A Note on the Concept of Delinking', Review, X/3 (1987), 435–444.

to exchange).[35] The political experiences that serve as reference points to these claims, such as the 1994 Zapatista insurrection in Mexico, aimed indeed at the production of autonomous localities disentangled from the neoliberal nation-state.[36] However crucial to a counter-cosmology these experiences might be, they cannot be considered sufficient solutions in themselves, at least if one does not want to lose sight of the planetary problem altogether and fall back into the local vs. global dichotomy—the fatal misbelief "that the global entails homogenization and undifferentiated identity whereas the local preserves heterogeneity and difference".[37] From a Western viewpoint, the specters of nativism and primitivism seem inescapable.

Without taking anything away from its fundamental strategic function, the decolonial discourse remains stuck into a rabbit-duck dynamic: if it draws the consequences from its own principles, it reduces itself to a localism; if it tries to integrate a more-than-localist discourse, it slips back into the old Western conceptual device of unity-in-plurality ("the planet is differentiated and yet one"[38]). This is what happens to the "pluriverse", itself inspired by the Zapatista battle cries. The pluriverse is the key concept of a political cosmology that strives to go beyond the planetary and towards "a world in which many worlds fit",[39] a web of local cosmologies that never subsumes them into an overarching totality. The fact that the term "pluriverse" was first used by Carl Schmitt to designate the political nature of

35 Anna L. Tsing, 'On Nonscalability: The Living World Is Not Amenable to Precision-Nested Scale', Common Knowledge, 18 (2012), 505–524; Id., 'Towards a Theory of Nonscalability', Multitudes, 82/1 (2021), 65–71.

36 See Alvaro Reyes, Mara Kaufman, 'Sovereignty, Indigeneity, Territory: Zapatista Autonomy and the New Practices of Decolonization", in Luisetti, Pickles and Kaiser, 44–68.

37 Toni Negri, Michael Hardt, *Empire* (Cambridge, MA, and London: Harvard University Press, 2000), pp. 44–45. On the political insufficiency of localism see also Greg Sharzer, *No Local: Why Small-Scale Alternatives Won't Change the World* (Winchester, UK and Washington: Zero Books, 2012).

38 Chakrabarty, p. 8.

39 De la Cadena, Blaser (eds.), *A World of Many Worlds*, p. 1.

the planet[40] is more than just an unacknowledged coincidence. In Schmitt, a Hobbesian premise justifies the order of the nation-state. Pluriversal thinkers reverse the polemological moral and identify it as a deeper truth beneath the mists of politics: this is the time of the "wars of the worlds"—where "war" is to be intended as "a conflict for which there is a no agreed-upon arbiter, a conflict in which what is at stake is precisely what is *common* in the common world to be built".[41] Horizontal symmetry means collision of viewpoints.

The positive side of this ecosystemic omni-conflictuality is the re-proliferation of indigenous *nomoi* or singular worlds, which includes practices of kinship and pact-making. Here another German ontology of the 1930s is inadvertently evoked, that of Martin Heidegger.[42] The pluriverse is a plural cosmos emerging from the infinite operations of world-forming, or "worlding", in which every human and non-human collective consist. Haraway's understanding of the term "worlding" is employed to maintain that one does not 'have' a world, but rather inhabits a world by participating in its making.

---

40 "*Die politische Welt ist ein Pluriversum, kein Universum*", "the political world is a pluriverse, not a universe" (Carl Schmitt, *The Concept of the Political*, expanded edition, trans. by G. Schwab [Chicago and London: The University Chicago Press, 1996], p. 53).

41 Bruno Latour, 'Whose Cosmos, Which Cosmopolitics? Comments on the Peace Terms of Ulrich Beck', Common Knowledge, 10/3 (2004), 450–462 (455). Latour deals with Schmitt openly in 'Comment on a Dialogue by Carl Schmitt', where he writes: "Even if, at the time, Schmitt's position looked infinitely more criminal than the positions of the winners [of WW2], are we so sure today, now that the twentieth century is revealing its true nature, of the lasting asymmetry between the two? [...] To the crimes committed in the name of the *Grossraum* and *Lebensraum* might now be added those committed in the name of a flight away to the Globe". As Osborne comments, "this should give us some pause for thought about the territorial politics of this new planetarianism" (Osborne, p. 11). Another German arch-reactionary should also be mentioned among Latour's references in relation to his anti-modernism: Eric Voegelin (see Fluss and Frimm, p. 106). In general, "Latour seems sanguine in flirting with highly reactionary jargon (even as he often professes liberal, internationalist policy prescriptions)" (Fluss and Frimm, p. 145).

42 Except for Chakrabarty, who enthusiastically embraces Heideggerianism since *Provincializing Europe*, this filiation too is usually overlooked.

This is, however, little less than an extension of Heidegger's theory of the world minus his metaphysics. Cosmology is brought back to the sphere of existence, and existence to the sphere of culture. The combination of Schmittian polemology and Heideggerian ontological anthropology seems to lead to a situation in which many autonomous cultures exist, negotiate their boundaries and are often at war with each other. Both the idealistic relationism of the postcolonials and the romantic relationism of the decolonials are based on the horizontality of relations among symmetric localities: it all comes down to "relations between two or more cultural 'zones', brought together in a meeting place".[43]

Take one last illustrious example: Isabelle Stenger's cosmopolitics, which significantly contributed to the formation of the pluriversal and Gaian terminology. Stengers acknowledges that any counter-cosmology of the present has to start from the counter-cosmological events of our epoch, like the 1999 WTO protests in Seattle. "Becoming the child of an event: not being born again into innocence, but daring to inhabit the possible as such"[44]—this is the task. Consistently, cosmopolitics avoids any universalization, abstraction or utopia: "Cosmos is always an emergent condition resulting from disagreements among divergent worlding practices participating in the discussion".[45] Despite the right premises, the discourse steers again towards a fictional situation resembling the fantasy of a pre-modern, even pre-historic first nature: a ground of "equality" without "equivalence".[46] If, alternatively, this is the world that cosmopolitics aims to bring about, then it is a post-apocalyptic world in which planetary-scale catastrophes have reset history. Either case, it is not the world of the present.

There is more to the dream of a planetarization 'from below'. Here my only purpose is to expose the theoretical and analytical rarefaction of what are considered some of the most progressive strains of critical theory today. Rarely do these proposals engage

---

43 Glissant, p. 14.

44 Philippe Pignarre and Isabelle Stengers, *Capitalist Sorcery: Breaking the Spell*, trans. by A. Goffey (London: Palgrave MacMillan, 2011), p. 4.

45 De la Cadena, Blaser (eds.), *A World of Many Worlds*, p. 12.

46 Stengers, 'The Cosmopolitical Proposal', p. 996.

seriously with sharper critical hypotheses like the Capitalocene, or with the social and economic stratifications of our epoch altogether; in fact, they seem to cut ties with the present and remain entrenched behind the ecological paradigm, which alone does not provide sufficient instruments for interpreting the state of things. To the silicium-cold abstractions and "necrophiliac *jouissance*"[47] of accelerationism, the Gaians, the postcolonials and the decolonials alike oppose a pathic thought and an ontology of touch among local *potentiae*. Whereas the *nomos* represents a principle of subjectivation that stabilizes an interiority against an exteriority, the principle of *pathos* is topologically reversible: a simultaneously active and passive, interior and exterior zone of co-presence that lies at the roots of differentiation and organization. This idea, however, is fundamentally at odds with the principle of autonomy and opacity of all terms involved. A relation is never really prior to its terms without—so to speak—negating their right of existence; the encounter is never really innocent. Hence the polemological realism and the voluntarist hope in a pact or a creative agreement.

The contradictions of ecological relationism are well epitomized by the paradoxical concept of "translation". Glissant exemplarily argued that like the literary word, the expressions of human collectives are untranslatable: opaque rather than transparent, concrete rather than general. This opaque concreteness is not mythological in that it does not conceal and naturalize existing power relations, but preserves "the Diverse".[48] Yet, human expressions are also intrinsically relational, and like words they can be translated. A reflexive gesture of commensurability ensures a universal *métissage*, a postmodern Baroque of glimmering identities. Today, the "aesthetics of *chaos-monde*"[49] is at the basis of the field of World Literature and the so-called translational turn. As it has been noted, "with translation assumed to be a good thing *en soi*—under the assumption that it is a critical praxis enabling communication across languages, cultures, time periods

47 Danowski, De Castro, p. 50.
48 Édouard Glissant, *Poetics of Relation*, trans. by B. Wing (Ann Arbor: University of Michigan Press, 2010), p. 62.
49 Glissant, *Poetics of Relation*, p. 94.

and disciplines—the right to the Untranslatable was blindsided. In a parallel way, at its very core World Literature seemed oblivious to the Untranslatable—as shown by its unqueried inclusion of the word 'world'".[50] Either translation (and therefore the institution of a meta-code) or opacity—*tertium non datur*.

If postcolonial aesthetics tends towards these new forms of planetary translatability and the contradictions of cultural identity, decolonial aesthetics is naturally inclined towards the untranslatable. Post-situationist *dérives* as flight strategies, performances materializing the spatial and temporal displacement of minorities, choreographies manifesting the political nature of body and space[51] are all decolonial gestures of expressive subtraction to the planetary codification and ways of learning how to inhabit the planetarized ruins. Whereas postcolonial aesthetics is typically an aesthetics of the word, decolonial aesthetics tends to be a bodily and performative aesthetics that unfolds at a distinctively sub-cosmological level (with a frequent overlapping of body and territory[52]). In giving form to the *pathos* of presence, it remains foreign to the constructive sensibility to the more-than-local. In its highest expressions, it also accepts that locality has been de-naturalized once and for all and keeps sublimation at the bare minimum so as to unsettle the boundaries between the aesthetic and the extra-aesthetic.

3.3. *The imaginary planet*

All the contradictions we have seen so far—those internal to the Gaian front, those between accelerationists and Gaians, those between anti- and pro-system cosmologies—are kept together by an

50 Emily Apter, *Against World Literature: On the Politics of Untranslatability* (London and New York: Verso, 2013), p. 15.

51 See e.g., Rizvana Bradley, *Anteaesthetics: Black Aesthetics and the Critique of Form* (Stanford: Stanford University Press, 2023) and André Lepecki, 'Choreopolice and Choreopolitics: Or, the Task of a Dancer', The Drama Review, 57/4 (2013), 13–27.

52 Nick Clare and Victoria Habermehl, 'Terrestrial Bodies', Dialogues in Human Geography, 15/3 (2024), 465–468.

image that serves to avoid confronting the monstrous issues of the present: the planet. Accelerationists and Gaians alike acknowledge that we are in the middle of a new cosmological phase raising from the ashes of globalization. Although without constituting a separate cosmology, the planet has become a new structure of awareness.

> Less and less relevant to the twenty-first century, modern paradigms appear increasingly unable to predict, let alone adequately explain, the global operations of technologically enhanced finance capital, cosmopolitanism's struggle to reinvent itself from the ashes of post-empire Europe, and the risk environment brought about by the ever-escalating crises of world ecologies. [...] The critical- theoretical model of planetarity attempts a move away from the totalizing paradigm of modern-age globalization—and thus a critique or critical "completion" of globalism—as well as from the irony and hermeneutics of suspicion typical of what came to be known as postmodernism.[53]

Planetarity does not oppose or replace globalization, but "overwrites" it (to use Spivak's term[54]) as a "new imaginary" driven by ethical and material preoccupations.[55] The planet presents itself as a relationism that supposedly eschews the multicultural diorama of neoliberal cosmopolitanism, "reworks designations of the migrant beyond the nation-state, but also gives rise to a set of considerations for how a 'practical dialogics' might keep in play multiple modes of human inhabitation without having to resolve or synthesize these, or to render them within a relation of domination or subjugation".[56]

---

53 Amy J. Elias, Christian Moraru, 'Introduction: The Planetary Condition', in Id., *The Planetary Turn: Relationality and Geo-Aesthetics in the Twenty-First Century* (Evanston: Northwestern University Press, 2015), xi–xxxvii (p. xi).

54 See Gayatri Chakravorty Spivak, *Death of a Discipline* (New York: Columbia University Press, 2005), p. 72. In this sense, all the terms in the semantic area of the 'planet'—planetary, planetarity, planetary thinking, planetarization, Thompson's "planetization"—form a consistent constellation despite their different origins and theoretical intentions.

55 Elias and Moraru, p. xxv.

56 Jennifer Gabrys, 'Becoming Planetary', e-flux, October 2018 (https://www.e-flux.com/architecture/accumulation/217051/becoming-

Finally, the planet is seemingly capable of catalyze the human and social sciences transversally.[57]

It is telling that the collection edited by Amy J. Elias and Christian Moraru in 2015 and titled *The Planetary Turn* proposes the task of a new "geoculture", this time however rooted in aesthetic knowledge (and specifically in media and literary studies) rather than in epistemology.[58] The goal is now to overcome even the differences between accelerationists and Gaians through the complete aestheticization and moralization of the planetary question. Once and for all severed from its social, historical and material background, planetary thinking becomes a poetic atmosphere, a cultural sensibility and a call to reconfigure our sensorium.[59] We are back where it all started: it is once again imperative to complement the techno-scientific image provided by Earth system science with a cultural awareness inspired by a newfound spiritualism without gods, an ethics—or even a theology—of immanence that reintroduces faith.[60]

That planetarity is now a fully aestheticized narration is demonstrated by the fact that it repeats the watchwords of the past without referring to almost any social and political aspect of the present. In an article published on *e-flux* in 2020, Yuk Hui exemplarily depicts the planetary as the task to create diversity, overcome the nation-state and establish a 'republic of knowledge' revolving around a new culture of technology. "A new diplomacy must arrive", he writes, "an epistemological diplomacy grounded in the project of technodiversity. This new diplomacy is more

planetary).

57 See Nigel Clark and Bronislaw Szerszynski, *Planetary Social Thought: The Anthropocene Challenge to the Social Sciences* (Cambridge, UK, and Medford: Polity Press, 2021).

58 See also Christian Moraru, *Reading for the Planet: Toward a Geomethodology* (Ann Arbor: University of Michigan Press, 2015).

59 See e.g., Josephine Berry, *Planetary Realism: Art Against Apocalypse* (Berlin: Sternberg Press, 2025).

60 Karen Bray, Heather Eaton, Whitney Bauman (eds.), *Earthly Things: Immanence, New Materialisms, and Planetary Thinking* (New York: Fordham University Press, 2023) The common reference here is to Alfred North Whitehead, already a seminal author for Thompson.

likely to be initiated by knowledge producers and intellectuals than by diplomats".[61] Although explicitly embracing a Heideggerian vocabulary and positioning himself against accelerationism, Hui also recovers a healthy dose of pre-accelerationist techno-enthusiasm (from Simondon for example). The battle between accelerationists and Gaians is a thing of the past. Hui's book from 2024, *Machine and Sovereignty: For a Planetary Thinking*, further substantiates this intuition: planetary thinking is a new moral sensibility, a new "language of coexistence".[62]

The emptiness of these neo-ecumenic appeals and ruminations of the collective imagery is grounded on a fundamental deafness to what the present is and how it presents itself to us, also on the aesthetic level. Even the irreparable conflict between abstraction and concretization has been finally subsumed into the latest—and, in terms of academic and editorial markets, trendiest—cosmological image. This persuasiveness comes from the pathos that lies at the core of this image, simultaneously a non-rational call to the collective consciousness and a pathic understanding of co-existence that evokes the values of posthuman mysticism (compassion, imagination, encounter) to skip over the problems of mediation. Unlike in the immediate post-war period, this pathos struggles to align itself with any real constituent effort, but also lacks the theoretical tools to delineate a truly interesting framework of analysis. Pumped by a highly aestheticized and increasingly reactionary industry of

---

61 Yuk Hui, 'For a Planetary Thinking', e-flux, 114, December 2020 (https://www.e-flux.com/journal/114/366703/for-a-planetary-thinking).

62 Yuk Hui, *Machine and Sovereignty: For a Planetary Thinking* (Minneapolis and London: University of Minnesota Press, 2024), p. 9. The similarities internal the planetary discourse are sometimes staggering. Writes Hui: "To think planetarily, first of all, means thinking beyond the configuration of modern nation-states, which have not been able to move away from vicious economic and military competition; second, it means formulating a language of coexistence that will allow diverse people and species to live on the same planet; and third, developing a new framework that will enable us to go beyond the question of territory, respond to the current ecological crisis, and reverse the accelerated entropic process of the Anthropocene" (p. 1).

knowledge, the planetary discourse capitalizes on the ecological and technological imaginary to conceal an epochal impasse.

4.
PLANET AND POWER

4.1. *A*-nomos, *counter*-nomos

What is the planet, then? Once the field is cleared of the haze of planetary ideology, the infrastructures of power relations and the socio-economic stratifications defining the current cosmological phase come to light with more clarity.

Most agree that the power topology of the planet is one of ungovernability, and that the planet cannot be governed as such—at least without the nature of sovereign power itself changing significantly. Carl Schmitt's already mentioned remark that "the political world is a pluriverse, not a universe" is symptomatic of the idea that a unified space is inherently extraneous to the concept of the political. At the beginning of the 1950s, Schmitt famously commented on the nascent global infrastructures in relation to the problem of a planetary *nomos* to come.[1] The birth of the planet, he posited, would be political only as a new repartition of differences. Technology is dangerous precisely because it is an a-political power of homogenization. The political

---

1 So reads the famous passage on the "third *nomos* of the Earth": "Development of modern technology has robbed the sea of its elemental character. A new, third dimension – air-space – has become the force-field of human power and activity. Today, many believe that the whole world, our planet, is now only a landing field or an airport, a storehouse of raw materials, and a mother ship for travel in outer space. That certainly is fantastic. But it demonstrates the power with which the question of a new *nomos* of the earth is being posed" (Carl Schmitt, *The* Nomos *of the Earth in the Jus Publicum Europaeum*, trans. by G.L. Ulmen [New York: Telos Press, 2003], p. 354).

order to come would therefore have to rely on the extensions of political subjectivity, i.e., of the sovereign nation-states broadened into macro-regions (*Grossräume*) or super-states. The former Trotskyist and prophet of the Cold War Left James Burnham had delivered a similar prognosis already in 1941: global sovereignty "will be restricted to the few super-states" because capitalism does not tolerate "an anarchic situation in global politics".[2]

Schmitt's position transpired the shock of the Nazi debacle and a sense of asphyxiation in front of the victory of technical abstraction and political anomie, embodied by the two post-WW2 blocks. His understanding of planetary politics was animated by a distinctive technophobia, which he shared with the likes of Heidegger and Jünger, and a phobia of political chaos. Behind these interrelated fears lurked the specter of the a-*nomos*, the absence of *nomos*, which by definition remains beyond the limits of the political and defines it *ex negativo*. Planetary politics appeared to the old German jurist as this unreadable political beast coming after the Behemoth and the Leviathan: one that remains grounded on political multiplicity, but subordinates it to a field of techno-economic totality; one that still relies on territorial sovereignty, but is not itself a form of territorial sovereignty. It is the Ziz, the biblical beast of the air—an undecipherable figure that never entered the scene of Schmitt's thought. For him, the planet rather functioned as the limit of Western political teratology.

A few decades later, when the global order had settled, Gilles Deleuze and Félix Guattari provided a different account of planetary political topology. Planetary power, they argued, works as a rule of translation of matter into form enforced by a polycentric network of zones of sovereignty. This rule is the capitalist "metaeconomics",[3] which underlies the government of raw flows and quanta (populations, food, cities, matter, energy, production…) by hand of the power structures of industrialized societies. The capitalist rule is not political in a traditional sense: it rather provides the value and the orientation of political structures, functioning more as an "axiomatic" than as a transcendent law—not an abstract ideal, but

2 Burnham, pp. 175, 172.
3 Deleuze and Guattari, p. 456.

an "integral of the power functions".[4] Liberalism reduced politics to folklore by overwriting it with a new link between the value of action (the capital) and its operability.

The planet, then, is a cosmological configuration of capitalism. Deleuze and Guattari combined the analysis of emerging neoliberalism with a metaphysics of immanence, and concluded that—again—the planet has no ruler, only operative rules: since "an axiomatic is not at all a transcendent, autonomous, and decision-making power opposed to experimentation and intuition", "it is an absurdity to postulate a world supergovernment that makes the final decisions".[5] Unlike politics, the capitalist axiomatic is not a field of decisions, but a semiotic grid of operations: whereas politics translates "the undecidable" (matter) into the decidable (form), the global capital functions by *limiting* the undecidable. It does act directly on the social matter from within it, but its action is only negative, contentless. It is an immanent limit, a rule of exchange. Its effects are political (and not only technical-economical), but not in a determining sense.

Not only do Deleuze and Guattari differ from Schmitt in their opposite stance towards anomie; they also argue very clearly that planetary governance is indeed political, although in a different sense than traditional power formations. Further, they are explicit about the fact that the problem is not the sanctification of technics per se, but its capitalistic instrumentalization.

Importantly, the French duo warned against any tragic distinction between politics and non-politics, or *nomos* and anomie. It is indeed a typical immune reaction of the sovereign apparatus that of expelling its otherness and labelling as a privation of itself (a-*nomos*). Anomie does not exist: it is only the phantasmal background necessary to the emergence of authoritarianism (a fiction that authoritarianism itself

4 Félix Guattari, *Piano sul pianeta. Capitale mondiale integrato e globalizzazione* (Verona: ombre corte, 2002), cap. 1 (my translation). Differently from a "programmatic", an axiomatic "does not have a definitively established program; it is always susceptible, when faced with a new crisis or an unforeseen difficulty, to the invention of new functional axioms or their suppression" (p. 43).

5 Deleuze and Guattari, p. 461.

tactically encourages in certain phases). Planetary politics is in fact characterized by a complex relationship between power structures—world-regions, nation-states as well as more granular forms of governance—and the axiomatic that integrates them (the capital), between structuring powers on one hand and this anti-formal potency on the other, between an organic and an ideal function.

Besides the formal function of power structures and the ideal function of the capital, there is also that which is formalized according to the idea: the material-energetic function, corresponding to the transversal ensemble of productive forces that are not regulated by any Power.[6] It is the "masses" or the planetary "minorities". For Deleuze and Guattari, the masses distinguish themselves from both power structures and the capitalist axiomatic for relational and topological reasons. First, they do not need to *organize* themselves in order to consist: they rather assemble without a pre-determined order (be it a form or an axiom). The matter-energy of minorities is "defined as a non-denumerable set, however many elements it may have. What characterizes the nondenumerable is neither the set not its elements; rather, it is the *connection*, the 'and' produced between elements, between sets [...]. The minorities constitute 'fuzzy', nondenumerable, nonaxiomizable sets, in short, 'masses'".[7]

From a strictly topological viewpoint, instead, minorities are describable as "holes" on the organization surfaces established by power structures and capitalist axioms. Holes are an ambiguous figure in *A Thousand Plateaus*: they open up a different topological dimension and establish new pathways of communication, but can also always become the centres of new zones of organization (valves, passages for regulating the fluxes—the fascistization of the mass). They may disrupt a surface but also contribute to its organization, that

---

6 Deleuze and Guattari, p. 226.

7 Deleuze and Guattari, p. 470. As it is well known, Deleuze and Guattari were inspired by the T.E. Lawrence's theory of guerrilla, which opposes the Turkish empire as an "algebraical element" that works by formulating "known variables, fixed conditions, spaces and times, inorganic things" and the Arab partisan as a "thing intangible", "drifting about like a gas" (Thomas Edward Lawrence, *Seven Pillars of Wisdom. A Triumph* [New York: Doubleday & Co. 1935], pp. 194–95).

is the evolution of its metabolism (it is around zones of 'digestion' of matter into form that a power structure constitutes itself). This is the chthonian nature of the masses, that same "dimension of the abyss" which Schmitt attributed to the figure of the partisan.[8]

In this framework, the capital is not an ascetic potency of abstraction but a metastable function of expression that treats the bodies and their potencies without expressing itself directly into stable forms.[9] It is a typically modern device that replaces the geometry of power with a material ideality on the organic model. Precisely in this sense the post-war capitalistic unification of the planet is asynchronous to its political unification and can still function through old political structures. The concept of planetary minorities, on the other hand, refers to the postcolonial masses and the residual masses of the West (the "internal Third Worlds, internal Souths"[10]). Deleuze and Guattari's use of the term "Third World" reflects the faith of the French revolutionary Left of the time (the *ultragauche*) in the epochal resurfacing of social matter-energy from the decolonial process[11] and in the planetary masses as the germ of a "people yet to come".[12]

Again, Deleuze and Guattari were wary of any stark opposition between organization and anomie. The centrality of the masses for the construction of a different planet does not entail a spontaneism

---

8 Carl Schmitt, *Theory of the Partisan: Intermediate Commentary on the Concept of the Political*, trans. by Gary L. Ulmen (New York: Telos Press 2007), p. 69. Schmitt's partisan too introduces a dimension of depth to the traditional surface topology of the political; however, it remains closer to the figure of the "fascist hero", an absolutization of the political as enmity and war, that is—in Deleuze and Guattari's terms—an already fascistized mass.

9 The immanent nature of capitalism as a system "that is constantly overcoming its own limitations", according to Deleuze, was key to the vision of Marx himself (Gilles Deleuze, 'Control and Becoming', in *Negotiations: 1972–1990*, trans. M. Joughin [New York: Columbia University Press, 1997], p. 171).

10 Deleuze and Guattari, p. 469.

11 See Alex Taek-Gwang Lee, *Communism After Deleuze* (London, New York, Oxford, New Delhi and Sydney: Bloomsbury, 2025), for example p. 9.

12 Deleuze and Guattari, p. 345.

of the revolutionary action, an eruption of anomie:[13] the masses must be mobilized by an act of unification, one that does not consist in a synthesis ("a synthesizer operating through rationalization, totalization, exclusion") but in an analysis ("an analyzer with respect to the desire of the group and the masses").[14] This is to say that the revolutionary machine is an a-subjective relational context that historically canalizes mass desire without subsuming it.

Schmitt and Deleuze-Guattari provide us with an abstract picture of the planetary *nomos* and its counter-*nomos*. The Schmittian Ziz is a post-political figure that remains fundamentally unthought, and in the three-quarters of a century after Schmitt's pronouncement has partially manifested as a polycentric globe and then a multipolar world. I will try to further describe this configuration in the next section. Deleuze and Guattari's characterization of the masses, on the other hand, aspires to define a planetary revolutionary subject. Such a subject is not the a-*nomos*, which is only a projection of the *nomos*, but rather a counter-*nomos*, a possible way to build another planet. Like the planetary *nomos* keeps together morality and power, the planetary counter-*nomos* must keep together the horizontality of desire and the verticality of a political machine capable of crossing different strata and mobilizing them against the *nomos*.

---

13 See Taek-Gwang Lee, pp. 70-71. "The problem", writes Deleuze, "is not at all about choosing between spontaneity and centralism. It serves no purpose to recognize in one breath the right to spontaneity during a first stage, if it means in the next breath demanding the necessity of centralization for a second stage: the theory of stages is the ruin of every revolutionary movement. From the start we have to be more centralist than the centralists. Clearly, a revolutionary machine cannot remain satisfied with local and occasional struggles: it has to be at the same time super-centralized and super-desiring. The problem, therefore, concerns the nature of unification, which must function in a transversal way, through multiplicity, and not in a vertical way, so apt to crush the multiplicity proper to desire" (Gilles Deleuze, 'Three Group Related Problems', in *Desert Island and Other Texts: 1953–1974*, ed. by D. Lapoujade, trans. by M. Taomina [Los Angeles: Semiotext(e), 2004], p. 199.

14 Deleuze, 'Three Group Related Problems', p. 198.

4.2. *The world recomposed*

System cosmology began after WW2 and culminated in the phenomenon called globalization, a phase already characterized by the saturation of planetary geography by the capital. "Since capitalism has invaded almost all exploitable economic surfaces, it cannot maintain the expansionary momentum that was characteristic during the colonial and imperialistic phases. This closure of its field of action requires it to recompose incessantly on itself, constantly reconverting its own economic and social spaces, its modes of control and subjection of the ensemble of human societies".[15] It goes without saying that globalization was not simply a geographic saturation (although it was not independent from geography), but first and foremost a psychosocial and economic invasion of life spaces that advanced in extensity as much as it penetrated in intensity. This is why even after becoming fully planetary, capitalism did not encountered any impassable limit—only the limits required for it to further transform. The only true limit of capitalism, as Deleuze had it, is itself.[16]

One of the main features of globalization has been its polycentrism. On the globalized planet, power structures maintained their old territorial nature, but they were also integrated into a centerless network whose complexity served the purpose of surviving local shocks. The illusion of globalization was a world in which "no one is in charge".[17] Yet of course, this neoliberal narration has always been aimed at naturalizing the international projection of the American empire and concealing the existential importance of institutions

---

15 Guattari, p. 46.

16 Deleuze, 'Control and Becoming', p. 171. The cannibalization of planetary capitalism seems to put an end to the development described by Arrighi, according to whom capitalism follows cycles of implantation, saturation and financialization. We might well be still in the third phase of the last cycle, but the germs of the next cycle appear inevitably more similar to a self-destructive action than to the conquest of new spaces (since a proper implantation in outer space is for the major part unachievable).

17 Milton Friedman, *The Lexus and the Olive Tree* (London: Picador, 2012), p.109.

for the 'free' market.[18] In this sense, the capital was a fundamental function of sovereignty that expanded the concept of sovereignty itself.

Since the end of the Cold War and manifestly since the 2007-2008 financial crisis, we find ourselves in a second subphase born out of the planetary dephasing. As a result of the saturation of physical and psychosocial geographies, capitalism delved deeper into its own post-industrial logic. Digital platforms added a further depth to the domain of extraction, production and reproduction and served as new instruments of crisis management; according to many, the combination of planetary financialization and digitalization represented a caesura in the traditional categorizations.[19] New biopolitical repartitions became necessary to justify the capitalist axiomatic—not least, a renewed call for global governance, as expressed for example in the Global Governance Forum 2024 (when the UN proposed the creation of an "Earth Council"[20]). Meanwhile, in the very last years, the determining function of international sovereign powers has rebecome visible around the Western centers: Israel and the US, Italy and France are drawing new social and physical borders and fueling internal and external diasporas. The "limiting" function of the capital no longer hides the determining action of the State, that of "*posing* the undecidable".[21] This action invests not only the permanent minorities, but more and

18 See Quinn Slobodian, *Globalists: The End of Empire and the Birth of Neoliberalism* (Cambridge, MA, and London: Harvard University Press, 2018).

19 See e.g., Andrea Fumagalli and Sandro Mezzadra (eds.), *Crisis in the Global Economy: Financial Markets, Social Struggles, and New Political Scenarios*, trans. by J.F. Mc Gimsey (Los Angeles: Semiotext(e), 2007).

20 See Global Governance Forum, '"*A Second United Nations Chart.*" *Modernizing the UN for a New Generation.* 2024/2025' (2024). The text is available at: https://globalgovernanceforum.org/second-united-nations-charter-modernizing-un-new-generation/.

21 Giorgio Agamben, *Homo Sacer: Sovereign Power and Bare Life*, in *The Omnibus Homo Sacer* (Stanford: Stanford University Press, 2017), p. 26 (my emphasis).

more also the impoverished petty bourgeoisie, exacerbating an erosive tendency that was already long at work.[22]

This cannibalistic development feeds on new ideologies of self-immunization. At a biopolitical level of analysis, we can draw on Povinelli's categories[23] to argue that "late-liberal capitalism" relies on a novel rearticulation of life and nonlife and therefore on new patterns of subjectivation and objectivation. While the modern biopolitical distinction grounded politics on the tragic separation of the organic (life) from the inorganic (nonlife), planetary biopolitics is now absorbing this separation into one and the same form of interiority (the organic-inorganic planet) and opposing it to a new otherness (nonlife as the non-planet). In other words, the distinction grounding politics and sovereignty is no longer that between *bios* and *geos*, but that between a planetary *bio-geos* and its cosmic alterity.

This allows to conceive of new subjectivation and objectivation devices at the ideological level. The planet itself becomes a form of domestic interiority that hosts 'humanity' as a whole, or the human regarded as a species, united in the face of common problems. The human is defined on the basis of its exceptionality, both internal (with respect to other species) and external (with respect to the inhuman non-planet). Outside the planet, a spatial and temporal otherness is conceived of as the new frontier: humanity is destined to conquer the non-human vastness of the universe and the non-human eternity of death (as predicated by the planetary ideologies described at the end of chapter 2). This large-scale endeavor of immunization is today an active effort of colonization, but also and above all a way to establish fictitious external limits in order to hide and justify internal operations of homogenization and further repartition. It is not hard to

---

22 "This deterritorialization of capitalism on itself, which Marx called 'expropriation of bourgeoisie by hand of the bourgeoisie', but that now is developing on an entirely different scale, does not imply that integrated global capital is universalist. It is not particularly fond of maintaining democratic bourgeoisie, as well as it is not inclined to a particular kind of dictatorship. The only thing that it strives to homogenize is the modes of production and the modes of social control" (Guattari, pp. 47-48).

23 Elizabeth Povinelli, *Geontologies: A Requiem to Late Capitalism* (New York: Duke University Press, 2016).

see, for instance, how the mythological amplification of the current discourse on extinction is weaponized in favor of neo-colonialist operations and exclusionary politics,[24] or how astrocapitalism opens new abstract economies that amplifies terrestrial inequalities.[25]

Ultimately, the immunization of the planet acts as a biopolitical justification of the existing forms of global governance, that is a transnational web of policies aimed at sustaining planetary capitalism. Efforts in the political and ethical legitimization of global governance are more and more common in and outside the academia, with a vast mobilization and production of concepts. For instance, the idea of "stewardship" (a variation of the idea of private property charged with Protestant morality) is often employed to designate a common responsibility to the planet which justifies meta-political and meta-state formations. Here ecologism becomes an opportunity for "market planetarianism".[26] Other proposals of world governance arise in anti-US function, as in the Chinese vision based on the imperial principle of *tianxia* ("all under one heaven"),

---

24 See e.g., Jim Igoe, Dan Brockington, 'Neoliberal Conservation: A Brief Introduction', Conservation and Society, 5/4 (2007), 432–449; Rosaleen Duffy, *Nature Crime: How We Are Getting Conservation Wrong*, (New Haven: Yale University Press, 2010); Dan Brockington, and Rosaleen Duffy (eds.), *Capitalism and Conservation* (London: Wiley-Blackwell, 2011).

25 See Peter L. Nelson, Walter E. Block, *Space Capitalism: How Humans Will Colonize Planets, Moons, and Asteroids* (Cham: Springer, 2018).

26 Fraser Murison Smith, *A Planetary Economy* (Cham: Palgrave Macmillan, 2020). "The difference between a planetarian society and today's", writes the author (an ecological economist with a career in consulting), "is that, in a planetarian one, markets will serve planetarian goals. They will constitute a means to an end rather than the end in itself, as they are considered under current market liberalism" (p. xxvvii). On environmental stewardship in industrial culture, see for instance Doug McKenzie-Mohr, Nancy R. Lee, P. Wesley Schultz and Philip Koetler, *Social Marketing to Protect the Environment: What Works* (SAGE, Thousand Oaks 2012). On the extension of this nexus into the notion of earth or planetary stewardship, see e.g., F. Stuart Chapin III, 'Transformative Earth Stewardship: Principles for Shaping a Sustainable Future for Nature and Society', Earth Stewardship 1/2024, e12023.

the "community of common destiny for mankind".[27] Such positions assume the neoliberal self-narration according to which globalization has been an age of unregulated individualism and argue that today we are in need of a less destructive and more democratic capitalism. "Guide the markets and let them do their work".[28]

Like globalization, planetarization acknowledges a plurality of international actors and encourages an a-centric configuration of national power structures. Globalized polycentricity is thus replaced by planetarized multipolarity, a versatile ideology shared by neofascist ideologues and neoliberal bureaucrats alike. Whereas polycentricity was the ideology of transparent relationality, multipolarity is a narrative of opacity, conflict and tension. Russian ideologue Aleksandr Dugin, for instance, makes it into the theoretical foundation of his metaphysical justification of Russia's imperial projection. "A multipolar world", he claims, "differs from the classical Westphalian system in that it does recognize in separate nation-states, legally and formally sovereign, the status of full-fledged poles", but does not replace it with a new "system of international relations that insists on the legal equality of nation-states, considered as matter-of-fact".[29] Dugin recovers Schmitt's polemological vision through an eclectic window-dressing criticism of capitalism, thoroughly associated with Western civilization.

Many authors across the whole political-philosophical spectrum regard multipolarilty as a chance to destitute Western hegemony,[30]

---

27 See for instance Tingyang Zhao, *Redefining a Philosophy for World Governance*, trans. by L. Tao (Singapore: Palgrave MacMillan, 2019), who writes: "Only by establishing a universal world order above and beyond the national system can the outdated imperialist hegemony be eliminated, the emerging new technological powers be contained and the world be saved from insanity and destruction" (p. 61). "No member in the system will gain maximized personal interests but can hope for maximized common security and shared interest", he explains (p. 73).

28 Murison Smith, p. 461.

29 Aleksandr G. Dugin, *The Theory of a Multipolar World* (London: Arktos, 2020), p. 8.

30 "If Westernization was unipolar, dewesternization is multipolar. Unipolarity was successful in enacting the global designs associated with Westernization. Multipolarity, on the other hand, can no longer be

regardless of the fact that it has already become a neoliberal buzzword. The theme of Davos 2022, multipolarity asks international managers to "build geopolitical muscle"—not in order to re-territorialize, but to further volatize: shape more resilient supply chains, decentralize and multiply, complete digitalization and generally "sense, anticipate, and respond to geopolitical shocks". It "isn't just about managing risk; it's about leveraging global shifts to create opportunities".[31] In sum, the idea of multipolarity conceals the restructuring of planetary capital behind an epochal shift away from the globalized pacification occurring on a geopolitical level: a "weaponized interdependence"[32] that does not hinder the "synarchic"[33] arrangements necessary to preserve and sustain the current modes of extraction and production.

4.3. *The alter-planet*

Globalization has been not only a deeply composite process, but also a largely contingent and accidental phenomenon, impossible to comprehend by neglecting the role of its opposers. In many ways, the international scope of anti-capitalist struggles "were the real motor that drove the development of the institutions of capital and that drove it in a process of reform and restructuring"[34] in a planetary sense. At the height of globalization, the composition of the anti-planetary

---

controlled by global designs; it fractures them, by definition" (Walter D. Mignolo, 'Foreward: On Pluriversality and Multipolarity', in Bernd Reiter [ed.], *Constructing the Pluriverse: The Geopolitics of Knowledge* [New York: Duke University Press, 2018], ix–xvi [p. xiii]).

31 Aparna Bharadwaj and Nikolaus Lang, 'Why Businesses Need "Geopolitical Muscle" in a Multipolar World', World Economic Forum (online), January 14, 2025 (https://www.weforum.org/stories/2025/01/businesses-need-geopolitical-muscle-multipolar-world/).

32 Henry Farrell and Abraham L. Newman, 'Weaponized Interdependence: How Global Economic Networks Shape State Coercion', International Security, 44 (2019), 49–79.

33 Daniel Woodley, *Globalization and Capitalist Geopolitics. Sovereignty and State Power in a Multipolar World* (London and New York: Routledge 2015), pp. 22ff.

34 Hardt and Negri, p. 51.

'movement of movements' was already so wide and diverse that the attempts to understand it and mobilize it as a whole raised difficult problems. Commenting on the 2001 World Social Forum in Porto Alegre, Michael Hardt famously spoke about "a swarming multitude and a network of movements", an "overflowing enormity", a "sea of people" that required a work of "mutual adequation".[35] At the turn of the century, building alliances and local alignments within this rich front seemed far more fruitful than aiming at an overarching counter-network.[36]

The alter-globalist cycle of struggles, started with the Zapatista insurrection in 1994 and culminated with the Genoa G8 in 2001, was the last large-scale political-cosmological laboratory of our epoch, and to that we need to go back in order to imagine a counter-*nomos* of the present. The movements of the squares could be included as "vanishing mediators—destructive operations which [...] clear the terrain for new and unexpected developments".[37] In the last 25 years, a less organized but no less vital constellation of planetary struggles has arisen, and our thinking must be attuned also to these experiences. Notoriously, these experiences leave open the question on political subjectivity. Whatever "the forms of coordination and organization of the future revolutionary movements" might be, prophesized Guattari, they would have to imply "the respect of autonomy and singularity of each of their components".[38]

Hardt and Negri have theorized a return from the "multitude" to the class understood as a "multitudinous class", that is an intersectionalist formation determined not by external associations (such as alliances and coalitions), but by internal bonds of solidarity and relations that transform their terms.[39] The question remains one of articulation—a

35 Michael Hardt, 'Today's Bandung?', New Left Review, 14 (2002), 112–118, here pp. 112–114.
36 Tom Mertes, 'Grass-Roots Globalism. Reply to Michael Hardt', New Left Review, 17 (2002), 101–110, here 108.
37 Frederic Jameson, 'The Aesthetics of Singularity', New Left Review, 92 (2015), 101–132 (p. 131).
38 Guattari, p. 58.
39 Michael Hardt and Antonio Negri, 'Empire, Twenty Years On', New Left Review, 120 (2019), 67–94 (p. 87).

question that, as notes Mezzadra, cannot be abstracted from a thorough analysis of how planetary capitalism itself articulates the social, political, psychical and physical strata.[40] Only after such analysis can we begin to bring into focus the morphing multiverse of a planetary class and its own "whole without closure".[41]

This book stops here, on the threshold of an alter-cosmology, which has been the vanishing point of my whole analysis. This other cosmology will have to be both transversal and infrastructural and rethink the relationism of postcolonial and post-patriarchal struggles in light of the verticality that technology allows. A rigorous techno-relationism built on a social and political critique of scale abstraction remains to be thought. To be sure, this aspect cannot but intertwine with the ecological problem and the issue of ecological subjectivation ("Who will build the ark?", asked Mike Davis[42]). Strategies for reappropriating the planetary commons will have to rely on analyses of how ecology today is harnessed for profit, exploitation and dispossession. At the core of all these problems there must be an analysis of the "murky megastructures"[43] of planetary labor, the flows of "live labor in transition"[44] and the related changes generated by automation.

Authors such as Simondon and Deleuze-Guattari did provide some theoretical elements to start rethinking scale and collective subjectivation. Despite all its acritical passion for planetary reticulations, Simondon lucidly identified the problem of reconnecting knowledge and praxis through technical infrastructure as the central problem of information age. Leaps in scale, he argued

40 Sandro Mezzadra, *La questione postcoloniale. Storia e politica nel presente globale* (Verona: ombre corte, 2008), p. 107.

41 Louis Althusser, 'Élements d'autocritique', in *Solitude de Machiavel et autres textes* (Paris: PUF, 1998), p. 184.

42 Mike Davis, 'Who Will Build the Ark?', New Left Review, 61 (2010), 29–46. "Since most of history's giant trees have already been cut down, a new Ark will have to be constructed out of the materials that a desperate humanity finds at hand in insurgent communities, pirate technologies, bootlegged media, rebel science and forgotten utopias" (p. 30).

43 Phil A. Neel, *Hellworld: The Human Species and the Planetary Factory* (Leiden and Boston: Brill, 2025), p. 30.

44 Mezzadra, pp. 120–123.

in the *Individuation*, are to be understood as genetic dynamics that combine spontaneity and organization to the point of their indistinguishability. Transindividual subjectivation itself is a passage in scale that passes through technicity and culture. The planetary should therefore be this work of reconciling knowledge and praxis through novel mediations at a transindividual level.

Analogously, Deleuze and Guattari highlighted the need for a reconceptualization of the 'mass' as the subject of the planetary counter-*nomos*. Masses should be conceived neither as the alienated opposite of the 'individuum' (the unproductive and undifferentiated chaos already deplored by Marxism) nor as a self-sufficient emancipatory dimension, as they rather evoke an altogether different grammar of collective subjectivation. The swarm-mass and the databank-mass are the other side of the contemporary figures of the "dividual" and the "Emigrant",[45] the main modes of subjectivation of information societies. The new collective combinations will have to be topologically ubiquitous, but also sensuous and affective. Such affectivity in the absence of organic structuration is unachievable outside the domain of technological intensity; intensity, not quantity, is the ground on which the battle around scale must be fought.

45 The "dividual" is mentioned in Gilles Deleuze, 'Postscript on the Societies of Control', October 59 (1992), 3–7, and the "Migrant" in Gilles Deleuze and Félix Guattari, *What is Philosophy?*, trans by H. Tomlinson and G. Burchell (New York: Columbia University Press, 1994), p. 98.

CONCLUSIONS
WHAT ARE WE SUPPOSED TO DO WITH THE PLANET?

Quei soldati bipedi come corrono guarda appesi alla bufera
ma cosa ce ne facciamo del pianeta!

Nanni Balestrini, *Ma dove stiamo andando…?*

"What are we supposed to do the planet!", exclaimed the Italian poet and militant Nanni Balestrini in 1961. Today this question has become deadly serious. We cannot do without the planet, and therefore we must ask ourselves what planet we want.

This book was a preliminary contribution to understanding what planet we do *not* want and what is wrong with our concepts of the planet. I have briefly investigated the historical roots of the current cosmology and identified the planet as its last subphase, emerging after WW2 and manifesting after the end of the Cold War. I have also analyzed its ideological ramifications into what I called the accelerationist and Gaian constellations, which represent its innermost contradiction and conceptual impasse. Stemming from the common strain of planetary culture, these trends introduce irrationalist cosmologies attempting at re-sacralizing reality with different means. Their relevance in contemporary debates lies in their fascinating effort of negating the wider systemic cosmology of our time (in their apparent 'rebelliousness'). Their efficacy in criticizing the present, however, is almost null, except for those fringes which can and must be co-opted away from their cosmological framework of reference.

Since this book is entirely possessed by the demon of criticism, it is perhaps important to underscore that some of the most radical thinkers and categories of accelerationism and Gaian ecologism can and should be taken as allies against what I defined as "pro-system" cosmological operations. No Manicheism is allowed in practice, and there surely are many elements from where one could begin to build—like Fisher's "acid communism", the ambiguous notion of "zone" and certain anarcho-indigenist positions. The point is that the cosmological coordinates of these perspectives are compromised in one sense or the other: either *plus ultra* or *plus intra*, to use Latour's terms.[1] Taking one way, however, should not mean refusing the other. The pathos of earthly touch should not exclude the a-corporeal intensities of technology, nor the construction of counter-platforms should neglect the forms of life in their concreteness, if we are to let new collective formations emerge that can confront planetary capitalism. This means both learning how to live among the planetarized ruins *and* inventing new devices to hack the noosphere.

This book was inspired by the conviction that the planet is not just an illusionary whole, and that a vital cosmology can and must be built today, as long as its social and political content is collectively determined first. At the same time, our planetary *Ganze* is not the minimum sufficient level for understanding the present. It is built, composed, and could be composed otherwise. It constitutes a milieu within which different social practices relate to and intertwine with each other.

In a cosmological perspective, aesthetics ceases to be an inert domain of freedom of the spirit or an equally inert reflection of the material base. Spiritualist and materialist heterodoxy either magnifies the aesthetic as the realm of free expression or reduce it to the sublimation of a truer, harsher reality. In fact, the aesthetic is *how* the base structure itself into sensible and symbolical regimes of experience. It is not at all the fine surface of a deeper stratum. From the standpoint of political cosmology, aesthetics is indeed a constitutive force that in itself is neither free and autonomously creative nor entirely complicit with the "mature alienation from human history and anthropocentric time and scale".[2] How can we be content with

1 Latour, *Facing Gaia*, p. 291.

2 Benjamin Bratton, 'Some Trace Effects, of the Post-Anthropocene: On Accelerationist Geopolitical Aesthetics', e-flux, 46 (2013), https://www.e-

the fact that "the intimacies with Algorithmic Capitalism" does not "provoke the prohibition of experiments"?[3] There is no such thing as a 'pure' constitution in the interstices of capitalism.

On the other hand, there is also no such thing as a pure pre-cognitive corporeality (or if there is, it is not particularly relevant). The aesthetic always involves thinking; creation is always also speculation—even as a force of singularization. The sensuous we have to look for is not the individual sensuous in a phenomenological sense, but the collective sensuous in the materialist sense of the word, that is after the determination of history. Precisely when it is not relegated into a separated domain of culture and is not passively received, when it is therefore a distributed force of social co-constitution, can aesthetics become an opportunity to not being complicit with the epoch and to understand its stakes in practice.

The aesthetic technologies of the alter-planet, be them postmodernist one-time devices[4] or permanent architectures on the modernist model, are not doomsday devices, devices of heavenly beauty, not even simple utopian devices—at least not right now. Their related cosmos is infinite and without closure, a-substantial, processual (a materialist "totality of the flow"[5]), but also internally structured across scales and domains of multiplicity. In this sense, aesthetic technology must become a privileged field of practice and experimentation if we want to get out of the hyper-aestheticized cosmological quicksand of our epoch.

---

flux.com/journal/46/60076/some-trace-effects-of-the-post-anthropocene-on-accelerationist-geopolitical-aesthetics.

3 Bratton, 'Some Trace Effects, of the Post-Anthropocene'.

4 See Jameson, 'The Aesthetics of Singularity', p. 113.

5 Boris Groys, *In the Flow* (London and New York: Verso, 2016), p. 10.

REFERENCES

Agamben, Giorgio, *Homo Sacer: Sovereign Power and Bare Life*, in *The Omnibus Homo Sacer* (Stanford: Stanford University Press, 2017).

Althusser, Louis, 'Élements d'autocritique', in *Solitude de Machiavel et autres textes* (Paris: PUF, 1998).

Amin, Samir, 'A Note on the Concept of Delinking', Review, X/3 (1987), 435–444.

Apter, Emily, *Against World Literature: On the Politics of Untranslatability* (London and New York: Verso, 2013).

Barbrook, Richard, *Imaginary Futures: From Thinking Machines to the Global Village* (London and Ann Arbor, MI: Pluto Press, 2007).

Barbrook, Richard, and Cameron, Andy, 'The Californian Ideology', in Richard Barbrook, *The Internet Revolution: From Dot-com Capitalism to Cybernetic Communism* (Amsterdam: Institute of Network Culture, 2015), 12–27.

Bardin, Andrea, *Hobbes' Materialist Agenda: The Politics of Early Modern Science* (Edinburgh: Edinburgh University Press, 2026).

Berger, Edmund, *Accelerazione. Correnti utopiche da Dada alla CCRU* (Roma: Nero, 2021).

Bergson, Henri, *The Two Sources of Morality and Religion*, trans. by R. Ashley Audra and Cloudesley Breton, with the assistance of W. Horsfall Carter (Westport, Conn.: Greenwood Press, 1974).

Berleant, Arnold, *Sensibility and Sense: The Aesthetic Transformation of the Human World* (Exeter: Imprint Academic, 2010).

Berry, Josephine, *Planetary Realism: Art Against Apocalypse* (Berlin: Sternberg Press, 2025).

Bharadwaj, Aparna, and Lang, Nikolaus 'Why Businesses Need "Geopolitical Muscle" in a Multipolar World', World Economic Forum (online), January 14, 2025 (https://www.weforum.org/stories/2025/01/businesses-need-geopolitical-muscle-multipolar-world/).

Bill, Max, and Sottsass, Ettore, 'Design and Theory', in Kathryn B. Hiesinger, George H. Marcus (eds.), *Design since 1945* (London: Thames and Hudson, 1983), 1–4.

Borsari, Andrea, 'On the Aesthetics of the Anthropocene: The Sublime and Beyond—Other Concepts and Forms of Visualization', European Journal of Creative Practices in Cities and Landscapes, 5/2 (2022), 242–258.

Bradley, Rizvana, *Anteaesthetics: Black Aesthetics and the Critique of Form* (Stanford: Stanford University Press, 2023).

Brand, Stewart (ed.), The Whole Earth Catalog. Access to Tools, Fall 1968.

Brand, Stewart, 'Why Haven't We Seen the Whole Earth?', in Lynda R. Obst (ed.), *The Sixties* (New York: Random House, 1977), 168–179.

Brand, Stewart, 'Rethinking Extinction', Aeon, April 21, 2015 (https://aeon.co/essays/we-are-not-edging-up-to-a-mass-extinction).

Brassier, Ray, *Nihil Unbound: Enlightenment and Extinction* (Palgrave Macmillan, 2007).

Bratton, 'Some Trace Effects of the Post-Anthropocene: On Accelerationist Geopolitical Aesthetics', e-flux, 46 (2013) (https://www.e-flux.com/journal/46/60076/some-trace-effects-of-the-post-anthropocene-on-accelerationist-geopolitical-aesthetics).

Bratton, Benjamin, *The Terraforming* (Moscow: Strelka Press, 2019).

Bray, Karen, Eaton, Heather, and Bauman, Whitney (eds.), *Earthly Things: Immanence, New Materialisms, and Planetary Thinking* (New York: Fordham University Press, 2023).

Brockington, Dan, and Duffy, Rosaleen (eds.), *Capitalism and Conservation* (London: Wiley-Blackwell, 2011).

Brzezinski, Zbigniew, *Between Two Ages: America's Role in the Technetronic Era* (New York: The Viking Press, 1970).

Burnham, James, *The Managerial Revolution: What is Happening in the World* (New York: John Day Company, 1941).

Capra, Fritjof, and Steindl-Rast, David, with Matus, Thomas, *Belonging to the Universe. Explorations on the Frontiers of Science and Spirituality* (San Francisco: Harper San Francisco, 1991).

Carchia, Gianni, 'Le rovine della rappresentazione. Lettura della «Critica del Giudizio»', in *La legittimazione dell'arte* (Napoli: Guisa, 1982).

Césaire, Aimé, *Discourse on Colonialism*, trans. by Joan Pinkham (New York: Monthly Review Press, 1972).

Chakrabarty, Dipesh, *The Climate of History in a Planetary Age* (Chicago and London: University of Chicago Press, 2021).

Chakrabarty, Dipesh, *One Planet, Many Worlds: The Climate Parallax* (Waltham, MA: Brandeis University Press, 2023).

Châtelet, Gilles, *Figuring Space: Philosophy, Mathematics and Physics*, trans. by Robert Shore and Muriel Zagha (Dordrecht: Springer, 2000).

Ciolkovskij, Konstantin, *Циолковский К. Э. Избранные труды* (Moscow: Academy of Sciences of the Soviet Union, 1962).

Clare, Nick, and Habermehl, Victoria, 'Terrestrial Bodies', Dialogues in Human Geography, 15/3 (2024), 465–468.

Clark, Charles M.A., *Economic Theory and Natural Philosophy: The Search for the Natural Laws of Economy* (Aldershot and Brookfield: Edward Elgar, 1992).

Clark, Nigel, and Szerszynski, Bronislaw, *Planetary Social Thought: The Anthropocene Challenge to the Social Sciences* (Cambridge, UK, and Medford: Polity Press, 2021).

Coale, Samuel Chase, 'Conspiracy and Paranoia', in Inger H. Dalsgaard (ed.), *Thomas Pynchon in Context* (Cambridge: Cambridge University Press, 2019), 211–216.

Coale, Samuel Chase, *Paradigms of Paranoia: The Culture of Conspiracy in Contemporary American Fiction* (Tuscaloosa: University of Alabama Press, 2005).

Congar, Yves, *Ecumenism and the Future of the Church* (Chicago: The Priority Press, 1967).

Danowski, Déborah, and Viveiros de Castro, Eduardo, *The Ends of the World* (Cambridge, UK, and Malden, US: Polity Press, 2016).

Davis, Erik, *Techgnosis: Myth, Magic and Mysticism in the Age of Information* (New York: Crown Publishers, 2004).

Davis, Mike, 'Who Will Build the Ark?', New Left Review, 61 (2010), 29–46.

de Fontenelle, Bernard Le Bovier, *Conversations on the Plurality of Worlds*, trans. by H.A. Hargreaves (Berkeley, Los Angels and Oxford: University of California Press, 1990).

de La Cadena, Marisol, and Blaser, Mario (eds.), *A World of Many Worlds* (Durham: Duke University Press, 2018).

de Luca, Tiago, *Planetary Cinema: Film, Media and the Earth* (Amsterdam: Amsterdam University Press, 2022).

de Rosnay, Joël, *The Macroscope: A New World Scientific System*, trans. by Robert Edwards (New York, Hagerstown, San Francisco, and London: Harper & Row, 1979).

de Rosnay, Joël, *Le cerveau planetaire* (Paris: Olivier Orban, 1986).

Deleuze, Gilles, 'Postscript on the Societies of Control', October 59 (1992), 3–7.

Deleuze, Gilles, 'Control and Becoming', in *Negotiations: 1972–1990*, trans. by Martin Joughin (New York: Columbia University Press, 1997) 169–175.

Deleuze, Gilles, 'Three Group Related Problems', in *Desert Island and Other Texts: 1953–1974*, ed. by David Lapoujade, trans. by Michael Taomina (Los Angeles: Semiotext(e), 2004), 193–203.

Deleuze, Gilles, and Guattari, Félix, *A Thousand Plateaus: Capitalism and Schizofrenia II*, trans. by Brian Massumi (Minneapolis and London: University of Minnesota Press, 1987).

Deleuze, Gilles, and Guattari, Félix, *What is Philosophy?*, trans. by Hugh Tomlinson and Grant Burchell (New York: Columbia University Press, 1994).

Demos, T.J., *Radical Futurisms: Ecologies of Collapse, Chronopolitics, and Justice-to-Come* (Berlin: Sternberg Press, 2023).

Derrida, Jacques, *The Beast and the Sovereign*, vol. 2, ed. by Michel Lisse, Marie-Louise Mallet and Ginette Michaud, trans. by Geoffrey Bennington (Chicago and London: The University of Chicago Press, 2011).

Descartes, René, *The World*, in *The World and Other Writings*, ed. by Stephen Gaukroger (Cambridge, UK: Cambridge University Press, 2004).

Duffy, Rosaleen, *Nature Crime: How We Are Getting Conservation Wrong*, (New Haven: Yale University Press, 2010).

Dugin, Aleksandr G., *The Theory of a Multipolar World* (London: Arktos, 2020).

Elias, Amy J., Moraru, Christian, Introduction: The Planetary Condition, in *The Planetary Turn. Relationality and Geo-Aesthetics in the Twenty-First Century* (Evanston: Northwestern University Press, 2015), xi–xxxvii (p. xi).

Engels, Frederick, *Dialectics of Nature*, trans. and ed. by Clemens Dutt, with a preface and notes by John B.S. Haldane (New York: International Publishers, 1940).

Escobar, Arturo, 'Against Terricide: Envisiong Paths Towards Pluriversal Transitions', Cultural Dynamics, 37/3 (2025), 236–244.

Faggioli, Massimo, and Froehle, Bryan, *Global Catholicism: Between Disruption and Encounter* (Leiden and Boston: Brill, 2025).

Falk, Richard A., 'Contending Approaches to World Order', Journal of International Affairs, 31/2 (1977), 171–198.

Farrell, Henry, and Newman, Abraham L., 'Weaponized Interdependence: How Global Economic Networks Shape State Coercion', International Security, 44 (2019), 49–79.

Feuerbach, Ludwig, *The Essence of Christianity*, trans. by Marian Evans (London: Trübner & Co., 1881[2]).

Fisher, Mark, 'Left Hyperstition 2: Be Unrealistic, Change What's Possible', K-punk, November 29, 2005 (https://k-punk.org/left-hyperstition-2-be-unrealistic-change-whats-possible/).

Fluss, Harrison, and Frim, Landon, *Prometheus and Gaia. Technology, Ecology and Anti-Humanism* (London and New York: Anthem Press, 2022).

Frelik, Paweł, 'Stanisław Lem's *Summa Technologiae* as Impossible Utopia', Science Fiction Studies, 40/3 (2013), 439–450.

Freudenthal, Gideon, and MacLaughlin, Peter (eds.), *The Social and Economic Roots of the Scientific Revolution* (Springer, 2009).

Friedman, Milton, *The Lexus and the Olive Tree* (London: Picador 2012).

Fuller, Richard Buckminster, *No More Secondhand God and Other Writings* (London and Amsterdam: Feffer & Simons, 1967).

Fumagalli, Andrea, and Mezzadra, Sandro (eds.), *Crisis in the Global Economy: Financial Markets, Social Struggles, and New Political Scenarios*, trans. by Jason Francis Mc Gimsey (Los Angeles: Semiotext(e), 2007).

Gebru, Timnit, and Torres, Émile P., 'The TESCREAL Bundle: Eugenics and the Promise of Utopia through Artificial Intelligence', First Monday, 29/4 (2024), https://doi.org/10.5210/fm.v29i4.13636.

Gheoghegan, Bernard D., *Code: From Information Theory to French Theory* (New York: Duke University Press, 2023).

Glissant, Édouard, *Poetics of Relation*, trans. by Betsy Wing (Ann Arbor: University of Michigan Press, 2010).

Glissant, Édouard, *Treatise on the Whole-World*, trans. by Cecilia Britton (Liverpool: Liverpool University Press, 2020).

Global Governance Forum, '"*A Second United Nations Chart.*" *Modernizing the UN for a New Generation.* 2024/2025' (2024).

Gorz, André, *The Immaterial: Knowledge, Value Capital*, trans. by Chris Turner (London and New York: Seagull Books, 2010).

Groys, Boris, *In the Flow* (London and New York: Verso, 2016).

Guattari, Félix, *Piano sul pianeta. Capitale mondiale integrato e globalizzazione* (Verona: ombre corte, 2002),

Hallward, Peter, *Out of This World: Deleuze and the Philosophy of Creation* (London: Verso, 2006).

Haraway, Donna, *Staying with the Trouble: Making Kin in the Chthulucene* (Durham: Duke University Press, 2016).
Hardt, Michael, 'Today's Bandung?', New Left Review, 14 (2002), 112–118.
Hardt, Michael, and Negri, Antonio, 'Empire, Twenty Years On', New Left Review, 120 (2019), 67–94.
Havers, Grant, 'The Right-Wing Postmodernism of Marshall McLuhan', Media, Culture & Society, 25 (2003), 511–525.
Hegel, Georg Wilhelm Friedrich, *Philosophy of Nature: Part Two of Encyclopedia of the Philosophical Sciences (1830)*, trans. by Arnold Vincent Miller (Oxford: Clarendon Press, 2004).
Heidegger, Martin, *On the Essence and Concept of φύσις in Aristotle's* Physics *B, I*, trans. by Thomas Sheehan, in *Pathmarks*, 183–230 (Cambridge: Cambridge University Press, 2013).
Hernstein-Smith, Barbara, 'Bruno Latour's Climate Evangelism', New Formations, 107–108 (2022), 230–235.
Hoffman, Stanley, 'Report of the Conference on Conditions of World-Order – June 12–19 1965, Villa Serbelloni, Bellagio, Italy", Dedalus, Spring 1966, 455–478.
Hoffman, Stanley, *Primacy or World Order: American Foreign Policy since the Cold War* (New York: McGraw-Hill, 1978).
Hofstaedter, Richard, *The Paranoid Style in American Style and Other Essays* (Cambridge, MA: Harvard University Press, 1996).
Hoskin, Michael, *William Herschel and the Construction of the Heavens* (London: Oldbourne, 1963).
Hui, Yuk, 'For a Planetary Thinking', e-flux, 114, December 2020 (https://www.e-flux.com/journal/114/366703/for-a-planetary-thinking).
Hui, Yuk, *Machine and Sovereignty: For a Planetary Thinking* (Minneapolis and London: University of Minnesota Press, 2024).
Hutchins, Robert Maynard, *St. Thomas and the World State* (Milwaukee, WI: Marquette University Press, 1949).
Igoe, Jim, and Brockington, Dan, 'Neoliberal Conservation: A Brief Introduction', Conservation and Society, 5/4 (2007), 432–449.
Jameson, Frederic, *The Geopolitical Aesthetics: Cinema and Space in the World System* (Bloomington and Indianapolis: Indiana University Press, 1992).
Jameson, Frederic, 'The Aesthetics of Singularity', New Left Review, 92 (2015), 101–132
Jameson, Frederic, 'Gherman's Anti-Aesthetic', New Left Review, 97 (2016), 95–105.

Kant, Immanuel, *Physische Geographie*, Erster Band, in *Kants Gesammelte Schriften*, Erste Abt., Band IX, hrsg. von der königlich Preußische Akademie der Wissenschaften (Berlin and Leipzig: De Gruyter, 1923).

Kant, Immanuel, *Critique of Pure Reason*, trans. and ed. by Paul Guyer and Allen W. Wood (Cambridge, UK: Cambridge University Press, 1998).

Katz, Michael, Marsh, William P., and Thompson, Gail Gordon (eds.), *Earth's Answer. Explorations of the Planetary Culture at the Lindisfarne Conferences* (New York: The Lindisfarne Association / Harper & Row, 1977).

Kauffman, Stuart A., *Reinventing the Sacred: A New View of Science, Reason, and Religion. The Science of Complexity and the Emergence of a Natural Divinity* (New York: Basic Books, 2010).

Kelly, Kevin, *Out of Control: The New Biology of Machines* (London: Fourth Estate, 1994).

Kissinger, Henri, *World Order: Reflections on the Character of Nations and the Course of History* (New York: Penguin Books, 2014).

Kothari, Ashish, Salleh, Ariel, Escobar, Arturo, Demaria, Federico, and Acosta, Alberto (eds.), *Pluriverse: A Post-Development Dictionary* (New Delhi: Tulika Books, 2019).

Lacan, Jacques, *The Seminar: Book XX. Encore 1972–1973*, ed. by Jacques-Alain Miller, trans. by Bruce Fink (New York and London: W.W. Norton & Co., 1998).

Lambert, Johann Heinrich, *Cosmologische Briefe über die Einrichtung des Weltbaues* (Augsburg: Eberhard Kletts Wittib., 1771).

Land, Nick, *The Thirst for Annihilation: Georges Bataille and Virulent Nihilism* (London: Routledge, 1992).

Land, Nick, 'Making it with Death: Remarks on Thanatos and Desiring-Production', in *Fanged Noumena. Collected Writings 1987–2007*, ed. by Robert Mackay and Ray Brassier (Falmouth: Urbanomic, 2012), 261–288.

Laplace, Pierre-Simon, *Philosophical Essays on Probabilities*, trans. by Andrew I. Dale (New York: Springer, 1995).

Latour, Bruno, 'Whose Cosmos, Which Cosmopolitics? Comments on the Peace Terms of Ulrich Beck', Common Knowledge, 10/3 (2004), 450–462.

Latour, Bruno, *Facing Gaia: Eight Lectures on the New Climatic Regime*, trans. by Catherine Porter (Cambridge, UK, and Madford, MA: Polity Press, 2017).

Latour, Bruno, *Down to Earth. Politics in the New Climatic Regime*, trans. by Catherine Porter (Cambridge, UK, and Madford, MA: Polity Press, 2018).

Latour, Bruno, and Chakrabarty, Dipesh, 'Conflicts of Planetary Proportions. A Conversation', Journal of the Philosophy of History, 14 (2020), 419–454.

Lazzarato, Maurizio, 'On the Californian Utopia/Ideology', in Diederich Diederichsen, Anselm Franke (eds.), *The Whole Earth: California and the Disappearance of the Outside* (Berlin: Sternberg Press, 2013), 166–168.

Lem, Stanisław, *Summa Technologiae*, trans. by Johanna Zylinska (Minneapolis and London: University of Minnesota Press, 2013).

Lepecki, André, 'Choreopolice and Choreopolitics: Or, the Task of a Dancer', The Drama Review, 57/4 (2013), 13–27.

Leroi-Gourhan, André, *Gesture and Speech*, trans. by Anna Bostock Berger (Cambridge, MA, and London: MIT Press, 1994).

Lévy, Pierre, 'Plissé fractal ou comment les machines de Guattari peuvent nous aider á penserle transcendental aujourd'hui', Chiméres, 8/21 (1994), 167–180.

Lovelock, James, *Gaia: A New Look at Life on Earth* (Oxford: Oxford University Press, 2000).

Luisetti, Federico, *Una vita. Pensiero selvaggio e filosofia dell'intensità* (Milano and Udine: Mimesis, 2011).

Luisetti, Federico, 'Demons of the Anthropocene. Facing Bruno Latour's Gaia', Philosophy Kitchen, 5/3 (2016), 175–186.

Luisetti, Federico, 'Decolonizing Gaia, or, Why the Savages Shall Fear Bruno Latour's Animism', Azimuth, 9 (2017), 61–70.

Luisetti, Federico, John Pickles, and Wilson Kaiser (eds.), *The Anomie of the Earth. Philosophy, Politics, and Autonomy in Europe and the Americas* (Durham and London: Duke University Press, 2015).

Lyotard, François, 'Can Thought Go on without a Body?', in *The Inhuman. Reflections on Time*, trans. by Geoffrey Bennington and Rachel Bowlby (Stanford: Stanford University Press, 1991).

Mackay, Robin, Pendrell, Luke, and Trafford, James (eds.), *Speculative aesthetics* (Falmouth: Urbanomic, 2015)

Marx, Karl, *Economic and Philosophic Manuscripts of 1844 and the Communist Manifesto*, trans. by Martin Milligan (New York: Prometheus Book, 1988).

Mbembe, Achille, *Out of the Dark Night: Essays on Decolonization* (New York: Columbia University Press, 2021).

Mbembe, Achille, *The Earthly Community: Reflections on the Last Utopia*, trans. by Steven Corcoran [Rotterdam: V2_, 2022].

McLuhan, Marshall, *Letters* (New York: Oxford University Press, 1987).

McLuhan, Marshall, 'Cybernetics and Human Culture' (1964), in *Understanding Me: Lectures and Interviews*, ed. by Stephanie McLuhan and David Staines (Toronto, 2003), pp. 44–55.

McLuhan, Marshall, and Powers, Bruce R., *The Global Village: Transformations in World Life and Media in the 21st Century* (New York: Oxford University Press, 1989).

McKenzie-Mohr, Doug, Lee, Nancy R., Schultz, P. Wesley, and Koetler, Philip, *Social Marketing to Protect the Environment: What Works* (SAGE, Thousand Oaks 2012).

Meillassoux, Quentin, *After Finitude: An Essay on the Necessity of Contingency*, trans. by Ray Brassier (London and New York: Continuum, 2008).

Mendlovitz, Saul H., 'The Program of the Institute of World Order', Journal of International Affairs, 31/2 (1977), 259–265.

Merleau-Ponty, Jacques, and Morando, Bruno, *The Rebirth of Cosmology* (New York: Knopf, 1976).

Merleau-Ponty, Maurice, *The Visible and the Invisible*, ed. by Claude Lefort, trans. by Alphonso Lingis (Evanston: Northwestern University Press, 1968).

Merleau-Ponty, Maurice, *Husserl at the Limits of Phenomenology: Including Texts by Edmund Husserl*, ed. by Leonard Lawlor with Bettina Bergo (Evanston: Northwestern University Press, 2002).

Mertes, Tom, 'Grass-Roots Globalism. Reply to Michael Hardt', New Left Review 17 (2002), 101–110.

Mezzadra, Sandro, *La questione postcoloniale. Storia e politica nel presente globale* (Verona: ombre corte, 2008).

Mignolo, Walter D., 'Foreward: On Pluriversality and Multipolarity', in Bernd Reiter [ed.], *Constructing the Pluriverse: The Geopolitics of Knowledge* (New York: Duke University Press, 2018), ix–xvi.

Minkowski, Eugène, *Vers une cosmologie. Fragments philosophiques* (Paris: Payot, 1999).

Moraru, Christian, *Reading for the Planet: Toward a Geomethodology* (Ann Arbor: University of Michigan Press, 2015).

Moravec, Hand, *Mind Children: The Future of Robots and Human Intelligence* (Cambridge, MA: Harvard University Press, 1988).

Moretto, Antonio, 'Negazione e libertà nella teoria hegeliana del meccanismo', Verifiche, XXVIII (1999), 3–23.

Murison Smith, Fraser, *A Planetary Economy* (Cham: Palgrave Macmillan, 2020).

Nancy, Jean-Luc, *The Sense of the World*, trans. by Jeffrey S. Librett (Minneapolis and London: University of Minnesota Press, 1997).
Neel, Philip A., *Hellworld: The Human Species and the Planetary Factory* (Leiden and Boston: Brill, 2025).
Negarestani, Reza, 'Globe of Revolution. An Afterthought on Geophilosophical Realism', Journal for Politics, Gender and Culture, 8/2 (2011), 25–54.
Negri, Antonio, 'Problemi di storia dello stato moderno', Rivista Critica di Storia della Filosofia, 22/2 (1967), 182–220.
Negri, Antonio, *The Political Descartes: Reason, Ideology and the Bourgeois Project*, trans. by Matteo Mandarini and Alberto Toscano (London and New York: Verso, 2007).
Negri, Antonio, and Hardt, Michael, *Empire* (Cambridge, MA, and London: Harvard University Press, 2000).
Nelson, Peter L., and Block, Walter E., *Space Capitalism: How Humans Will Colonize Planets, Moons, and Asteroids* (Cham: Springer, 2018).
Neyrat, Frédéric, *The Unconstructable Earth: An Ecology of Separation* (New York: Fordham University Press, 2019).
Osborne, Peter, 'The Planet as Political Subject?', New Left Review, 145 (2024), https://newleftreview.org/issues/ii145/articles/peter-osborne-the-planet-as-political-subject.
Oxman, Neri, 'Editorial: Material Ecology', Computer-Aided Design, 60 (2015), 1–2.
Oxman, Neri, 'Dermi-Domus: A Grown Wardrobe for Bodies and Buildings', Architectural Design, 87/6 (2017), 16–25.
Pignarre, Philippe, and Stengers, Isabelle, *Capitalist Sorcery: Breaking the Spell*, trans. by Andrew Goffey (London: Palgrave MacMillan, 2011).
Plant, Sadie, and Land, Nick, *Cyberpositive*, in Robin Mackay and Armen Avanessian (eds.), *#Accelerate#: The Accelerationist Reader* (Falmouth: Urbanomic, 2014), 303–313.
Povinelli, Elizabeth, *Geontologies: A Requiem to Late Capitalism* (New York: Duke University Press, 2016).
Prigogine, Ilya, and Stengers, Isabelle, *Order Out of Chaos: Man's New Dialogue with Nature* (London: Verso, 2018).
Rabo, Annika, 'Conspiracy Theory as Occult Cosmology in Anthropology', in Michael Butter and Peter Knight (eds.), *Routledge Handbook of Conspiracy Theories* (London and New York: Routledge, 2020), 81–93.
Reed, Patricia, 'What is Care at Planetary Dimensions?', lecture given at Floating University, Berlin, August 6, 2019 (https://laboriacuboniks.net/texts-interviews/what-is-care-at-planetary-dimensions/).

Schlanger, Nathan. 'Between Endorsement and Disavowal: André Leroi-Gourhan's Russian Interactions', Technology and Language, 5/2 (2024), 136–152.

Schmitt, Carl, *The Concept of the Political*, expanded edition, trans. by George Schwab (Chicago and London: The University Chicago Press, 1996).

Schneekloth, Lynda H., 'Unredeemably Utopian: Architecture and Making/Unmaking the World', Utopian Studies, 9/1 (1998), 1–25.

Scott, Felicity D., *Architecture or Techno-utopia: Politics After Modernism* (Cambridge, MA, and London: MIT Press, 2007).

Sessions, Hammond David, *Man, Machines, and Modernity: Inventing "Industrial Society" in French Sociology 1930–1981*, PhD Dissertation, Boston College, 2021.

Sgalambro, Manlio, *La morte del sole* (Milano: Adelphi, 2017).

Sharzer, Greg, *No Local: Why Small-Scale Alternatives Won't Change the World* (Winchester, UK, and Washington: Zero Books, 2012).

Shortall, Sarah, '"Building the Earth": Pierre Teilhard de Chardin, Science, and the Spirituality of the United Nations', Journal of the History of Ideas, 85/4 (2024), 827–855.

Simondon, Gilbert, *On the Mode of Existence of Technical Objects*, trans. by Cécile Malaspina and John Rogove (Minneapolis: Univocal, 2017).

Skirbekk, Gunnar, 'Bruno Latour's Anthropology of the Moderns: A Reply to Maniglier', Radical Philosophy, 189 (2015), 45–47.

Slobodian, Quinn, *Globalists: The End of Empire and the Birth of Neoliberalism* (Cambridge, MA, and London: Harvard University Press, 2018).

Sottsass, Ettore, 'The Planet as a Festival', Design Quarterly, 89 (1973), 7–16.

Spivak, Gayatri Chakravorty, 'Imperative to Re-Imagine the Planet', in *An Aesthetic Education in the Era of Globalization* (Harvard: Harvard University Press, 2012), 335–350.

Spivak, Gayatri Chakravorty, *Death of a Discipline* (New York: Columbia University Press, 2005).

Stengers, Isabelle, 'The Cosmopolitical Proposal', in Bruno Latour and Peter Weibel (eds.), *Making Things Public: Atmospheres of Democracy* (Cambridge, MA: MIT Press, 2005), 994–1003.

Stengers, Isabelle, *In Catastrophic Times: Resisting the Coming Barbarism* (Lüneburg: Meson Press, in collaboration with Open Humanities Press, 2015).

Stuart Chapin III, F., 'Transformative Earth Stewardship: Principles for Shaping a Sustainable Future for Nature and Society', Earth Stewardship, 1/2024, e12023.

Swedenborg, Emanuel, *Opera Philosophica et Mineralia*. Vol. I, *Principia Rerum Naturalium sive Novorum Tentaminum phenomena mundi elementaris philosophice explicandi* (Dresden and Leipzig: Friedrich Heckel, 1734).

Taek-Gwang Lee, Alex, *Communism After Deleuze* (London, New York, Oxford, New Delhi and Sydney: Bloomsbury, 2025).

Tafuri, Manfredo, *Architecture and Utopia: Design and Capitalist Development*, trans. by B.L. La Penta (Cambridge, Mass, and London: MIT Press).

Teilhard de Chardin, Pierre, *Human Energy*, trans. by J.M. Cohen (New York: Harcourt, 1969).

Teilhard de Chardin, Pierre, *Activation of Energy*, trans. by René Hague (New York: Harcourt, 1970).

Thacker, Eugene, *In the Dust of this Planet: Horror of Philosophy* (Winchester and Washington: Zer0 Books, 2011).

Thompson, William Irwin, *At the Edge of History: Speculations on the Transformation of Culture* (New York, Evanston, San Francisco and London: Harper & Row, 1972).

Thompson, William Irwin, *Passages About Earth: An Exploration of the New Planetary Culture* (New York, Evanston, San Francisco and London: Harper & Row, 1973).

Thompson, William Irwin, *Thinking Together at the Edge of History: A Memoir of the Lindisfarne Association, 1972-2012* (Traverse City, MI: Lorian Press, 2016).

Thompson, William Irwin, 'It's Already Begun: The Planetary Age is an Unacknowledged Reality', The New Story, Winter 1985–86, 26.

Thompson, William Irwin (ed.), *Gaia, a Way of Knowing: Political Implications of the New Biology* (Great Barrington, MA: Inner Traditions/Lindisfarne Press, 1987).

Tighe, Carl, 'Stanisław Lem: Socio-Political Sci-Fi', The Modern Language Review, 94/3 (1999), 758–774.

Tsing, Anna Lowenhaupt, 'On Nonscalability: The Living World Is Not Amenable to Precision-Nested Scale', Common Knowledge, 18 (2012), 505–524.

Tsing, Anna Lowenhaupt, 'Towards a Theory of Nonscalability', Multitudes, 82/1 (2021), 65–71.

Tsing, Anna Lowenhaupt, Bubandt, Nils, Gan, Elaine, and Swanson, Heather Anne (eds.), *Arts of Living on a Damaged Planet. Ghosts and Monsters of the Anthropocene* (Minneapolis and London: University of Minnesota Press, 2017).

Toscano, Alberto, and Jeff Kinkle, *Cartographies of the Absolute* (Winchester, UK, and Washington, USA: Zero Books, 2015).

Turner, Fred, *From Counterculture to Cyberculture: Stewart Brand, The Whole Earth Network, and the Rise of Digital Utopianism* (Chicago and London: The University of Chicago Press, 2006).

Veal, Demien, (ed.), *Collapse*, vol. V, *The Copernican Imperative* (Falmouth: Urbanomic, 2009).

Vidokle, Anton, 'Immortality and Resurrection for All!', in *Citizens of the Cosmos* (Berlin: Sternberg Press, 2024), 165–183.

Virno, Paolo, *L'idea di mondo. Intelletto pubblico e uso della vita* (Macerata: Quodlibet, 2015).

Wallerstein, Immanuel, *World-System Analysis: An Introduction* (Durham and London: Duke University Press, 2004).

Wallerstein, Immanuel, *The Modern World-System II: Mercantilism and the Consolidation of the European World-Economy, 1600–1750* (Berkeley, Los Angeles and London: California University Press, 2011^{2}).

Wallerstein, *The Modern World-System IV: Centrist Liberalism Triumphant, 1789–1914* (Berkeley, Los Angeles and London: University of California Press, 2011).

Watson, Matthew C., 'Cosmopolitics and the Subaltern: Problematizing Latour's Idea of the Commons', Theory, Culture & Society, 28/3 (2011), 55–79.

Wittgenstein, Ludwig, *Tractatus Logicus-Philosophicus* (London: Kegan Paul, Trench, Trubner & Co., 1922).

Woodard, Ben, *On an Ungrounded Earth: Toward a New Geophilosophy* (New York: Punctum, 2013).

Woodley, Daniel, *Globalization and Capitalist Geopolitics: Sovereignty and State Power in a Multipolar World* (London and New York: Routledge, 2015).

Woynarski, Lisa, *Ecodramaturgies: Theatre, Performance and Climate Change* (Cham: Palgrave MacMillan, 2020).

Zhao, Tingyang, *Redefining a Philosophy for World Governance*, trans. by Liqin Tao (Singapore: Palgrave MacMillan, 2019).

Zylinska, Joanna, *Nonhuman Photography* (Cambridge, MA: MIT Press, 2017).

MIMESIS GROUP
www.mimesis-group.com

MIMESIS INTERNATIONAL
www.mimesisinternational.com
info@mimesisinternational.com

MIMESIS EDIZIONI
www.mimesisedizioni.it
mimesis@mimesisedizioni.it

ÉDITIONS MIMÉSIS
www.editionsmimesis.fr
info@editionsmimesis.fr

MIMESIS COMMUNICATION
www.mim-c.net

MIMESIS EU
www.mim-eu.com

www.ingramcontent.com/pod-product-compliance
Lightning Source LLC
LaVergne TN
LVHW100732061126
851857LV00011B/503
9788869775529